Aroma Active Compounds in Foods

ACS SYMPOSIUM SERIES **794**

Aroma Active Compounds in Foods

Chemistry and Sensory Properties

Gary R. Takeoka, Editor
U.S. Department of Agriculture

Matthias Güntert, Editor
Haarman and Reimer GmbH.

Karl-Heinz Engel, Editor
Technische Universität München

American Chemical Society, Washington, DC

Library of Congress Cataloging-in-Publication Data

Aroma active compounds in foods : chemistry and sensory properties / Gary R. Takeoka, Matthias Güntert, Karl-Heinz Engel, editors.

p. cm.—(ACS symposium series ; 794)

Includes bibliographical references and index.

ISBN 0–8412–3694–1

1. Food—Odor—Congresses. 2. Flavor essences—Congresses. 3. Flavor—Congresses

I. Takeoka, Gary R. II. Güntert, Matthias. III. Engel, Karl-Heinz, 1954- IV. American Chemical Society, Division of Agricultural and Food Chemistry. V. American Chemiscal Society. Meeting (218th : 1999 : New Orleans, La.) VI. Series.

TP372.5 .A76 2001
664′.06—dc21 2001022725

The paper used in this publication meets the minimum requirements of American National Standard for Information Sciences—Permanence of Paper for Printed Library Materials, ANSI Z39.48–1984.

Distributed by Oxford University Press

PRINTED IN THE UNITED STATES OF AMERICA

Foreword

The ACS Symposium Series was first published in 1974 to provide a mechanism for publishing symposia quickly in book form. The purpose of the series is to publish timely, comprehensive books developed from ACS sponsored symposia based on current scientific research. Occasionally, books are developed from symposia sponsored by other organizations when the topic is of keen interest to the chemistry audience.

Before agreeing to publish a book, the proposed table of contents is reviewed for appropriate and comprehensive coverage and for interest to the audience. Some papers may be excluded to better focus the book; others may be added to provide comprehensiveness. When appropriate, overview or introductory chapters are added. Drafts of chapters are peer-reviewed prior to final acceptance or rejection, and manuscripts are prepared in camera-ready format.

As a rule, only original research papers and original review papers are included in the volumes. Verbatim reproductions of previously published papers are not accepted.

ACS Books Department

Dedication

We dedicate this book to our friend and mentor,
Dr. Roy Teranishi (1922–2000)

Contents

Correlation between Sensory Properties and Chemical Structures of Flavor Components

Synthetic, Thermal Reaction, and Enzymatic Approaches to Flavor Components

Additional Properties of Flavor Components and Flavors

Indexes

Preface

This book covers worldwide scientific results of flavor research that were presented in August 1999 at an American Chemical Society (ACS) symposium in New Orleans held in honor of Dr. Roy Teranishi's 77th birthday. Roy spent a great part of his career on flavor research. He has been one of the pioneers in this field and he definitely left his footprint in many publications and books published within the past 40 years. All of us, the three editors of this book and many of the speakers of this symposium, were influenced by him to some extent in their own careers dealing with flavors in foods. The main intent of the symposium and this book was consequently, to give Roy something back, because it had always been an honor and a pleasure to work with him. Unfortunately, the finalization of the volume was overshadowed by the death of Roy in December 2000. Due to this unexpected loss, this book turned into a last tribute to a true friend and an ideal.

Our symposium was organized to cover the newest trends in flavor research. Our intention was to invite as many international researchers as possible as well as to get a good mixture of speakers from industry and academia. Consequently, we had scientists from Asia (3), Europe (9), and from the United States (9), representing universities, research organizations, flavor companies, and food companies.

Over the years, the focus in flavor research has shifted remarkably from overall analytical investigations of different foods toward more specific research on individual flavor components that contribute to the sensory impression of a food. Not only the qualitative aspect counts but also the quantity in which the respective flavor component is present and its odor and taste thresholds. We learn more and more that despite the fact that thousands of volatile components have been characterized in foods during the past four decades only a few hundred of these may actually contribute sensorically to the flavor of different foods. Another interesting observation is that some of the recently analyzed flavor components are very potent and are only present in trace amounts (sub-ppb level) in foods. Therefore, qualitative and quantitative flavor analysis remains very challenging despite the fact that modern analytical equipment and correspondingly, data processing capabilities have improved dramatically.

The results presented at the symposium and here in the book reflect the latest status of flavor research. The data may be another piece on the way to fully understand how food flavors are constituted and work and thus help to continue the pioneering work started in this fascinating field many years ago by Dr. Roy Teranishi.

Acknowledgements

The symposium and book would not have been possible without the financial support of the ACS Division of Agricultural and Food Chemistry, Dragoco, Givaudan Roure, Haarmann & Reimer, T. Hasegawa & Company, Ltd., Takasago International Corp., and Dr. Tomoyuki Tsuneya. We are grateful for their generous contributions.

Gary R. Takeoka
Western Regional Research Center
Agricultural Research Service
U.S. Department of Agriculture
800 Buchanan Street
Albany, CA 94710

Matthias Güntert
Haarmann & Reimer
300 North Street
Teterboro, NJ 07608

Karl-Heinz Engel
Technische Universität München
Lehrstuhl für Allgemeine Lebensmitteltechnologie
Am Forum 2
D–85350 Freising-Weihenstephan
Germany

Chapter 1

Chemical and Sensory Characterization of Food Volatiles: An Overview

Karl-Heinz Engel

Technische Universität München, Lehrstuhl für Allgemeine Lebensmitteltechnologie, Am Forum 2, D–85350 Freising, Germany

Flavor research has continuously developed from the mere identification of volatiles to the detailed chemical characterization of aroma compounds and the assessment of their sensorial significance. Principles underlying the formation of volatile constituents are studied. The importance of ubiquitous phenomena in natural products such as chirality to flavor chemistry is recognized. Modern analytical techniques increasingly allow researchers to quantify the sensorial contribution of aroma compounds to food flavors.

Introduction

Over the past decades, flavor research has witnessed tremendous developments *(1, 2)*. The availability of coupled capillary gas chromatography – mass spectrometry as routinely applicable standard analytical tool initiated numerous research activities resulting in the identification of a complex spectrum of volatile food constituents *(3)*. However, only in recent years these investigations have evolved from the mere detection of volatile compounds to studies of other essential aspects of flavor chemistry. Examples of such areas are (i) the pathways underlying the formation of volatiles either by biogenesis or by thermal treatment, (ii) the chirality of many flavor compounds and the routes being developed for the synthesis of enantiomers by chemical procedures or by enzyme-catalyzed reactions, and eventually (iii) the sensorial significance of compounds studied by investigating the contribution of volatile constituents to the flavor of certain foods.

Formation of Volatiles

Two basic routes of flavor formation can be differentiated: biosynthesis of compounds via genetically determined pathways *(4, 5)* and thermally induced reactions resulting in volatile compounds *(6)*. In both areas the investigation of model systems starting from potential precursors or intermediates plays an important role *(7, 8)*. Studies on the formation of aroma compounds have especially profited from the application of isotopically labelled substances and the possibility to elucidate pathways by following their fate by GC/MS. The biosynthesis of key odorants, e.g. lactones *(9)*, as well as complex reaction sequences such as the Maillard reaction *(10)* have been followed on the basis of this approach.

Increasing knowledge has been accumulated in recent years on the liberation of aroma compounds from non-volatile precursors by acid or enzyme-catalyzed reactions *(11)*. The availability of LC/MS (MS/MS) will contribute to narrowing the gap between the knowledge on the spectrum of volatiles and the structures of the non-volatile conjugates *(12)*. In addition to the meanwhile "classical" monoterpene glycosides *(13)*, the role of precursors for other sensorially important classes of compounds, e.g. sulfur-containing volatiles, has recently been shown *(14)*.

Chirality of Aroma Compounds

The influence of the configuration of a chiral aroma compound and its sensory properties is a well-recognized principle which has been demonstrated for many examples *(15)*. Accordingly, the synthesis of optically pure enantiomers is of increasing importance. In addition to the classical procedure of chemical synthesis *(16)*, enzymes *(17)* or microorganisms *(18)* are increasingly applied.

Determination of the naturally occurring configurations of chiral volatiles, e.g. by capillary gas chromatography using chiral stationary phases, has contributed to the understanding of the biosynthesis of aroma compounds *(19)*. Together with isotope ratio analysis *(20)*, the determination of enantiomeric compositions has become an indispensable tool in authenticity assessment *(21)*.

Sensory Evaluations

Based on the concept of "aroma activity values" developed in 1960s *(22, 23)*, methods to determine the contribution of volatile constituents to the overall flavor of foods have been continuously refined. The following sequential procedure has been successfully applied to many examples: (i) identification of the most potent aroma contributors by aroma extract dilution analysis (AEDA) *(24)*, (ii) quantification of the essential aroma constituents using isotopically labelled standards *(25)*, (iii) determination of odor thresholds and "aroma activity values", and (iv) confirmation of the analytical data by sensory assessment of reconstituted mixtures *(26)*.

In addition to classical studies on structure-activity relationships *(15)*, correlation models based on physico-chemical properties are increasingly applied *(27)*.

Other Aspects

The trend towards "functional foods", i.e. foods with additional, health-promoting factors, is causing increasing reassessments of volatiles for effects going beyond the sensory properties. Anti-bacterial activities *(28)* or anti-oxidative effects *(29)* are examples for such long known phenomena now being investigated in more detail.

Future Perspectives

The need for detailed knowledge on the pathways underlying the formation of aroma compounds will increase. Improvement of the sensory properties of a food might be one of the areas suitable to demonstrate to the consumer potential benefits of modern techniques, such as genetic engineering *(30)*. This in turn requires knowledge of the regulatory mechanisms underlying flavor formation and the precise assignment of the key enzymes in a metabolic route to be influenced *(31)*.

Studies on the sensory properties will have to develop from a merely descriptive to predictive levels. The increasing capacity to make use of information on compounds will eventually enable the prediction of the sensory properties of molecules based on a comprehensive set of physical and chemical data processed via sophisticated and "intelligent" computer modelling approaches. By combining these approaches with more detailed studies on the physiology of flavor perception and the role of proteins and receptors involved in this process *(32, 33)*, the overall understanding of the role of chemical substances as flavor compounds should increase.

References

1. *Flavor Chemistry. Thirty Years of Progress*; Teranishi, R.; Wick, E.L.; Hornstein, I., Eds.; Kluwer Academic/Plenum Publishers: New York, 1999.
2. *Frontiers in Flavour Science*; Schieberle, P.; Engel, K.-H., Eds.; Deutsche Forschungsanstalt für Lebensmittelchemie: Garching, Germany, 2000.
3. *Volatile Compounds in Food: Qualitative and Quantitative Data*, 7th ed.; Nijssen, L. M., Ed.; TNO Nutrition and Food Research Institute, Zeist, Netherlands, 1996.

4. *Biotechnology for Improved Foods and Flavors*; Takeoka, G. R.; Teranishi, R.; Williams, P. J.; Kobayashi, A., Eds.; ACS Symposium Series 637; American Chemical Society: Washington, D.C., 1995.
5. Little, D. B.; Croteau, R. B. In *Flavor Chemistry. Thirty Years of Progress*; Teranishi, R.; Wick, E. L.; Hornstein, I., Eds.; Kluwer Academic/Plenum Publishers: New York, 1999; pp 239-253.
6. *Thermal Generation of Aromas;* Parliment, T.; McGorrin, R.; Ho, C. T., Eds.; ACS Symposium Series 409; American Chemical Society: Washington, D.C., 1989.
7. Williams, P. J.; Sefton, M. A.; Francis, I. L. In *Flavor Precursors. Thermal and Enzymatic Conversions*; Teranishi, R.; Takeoka, G. R.; Güntert, M., Eds.; ACS Symposium Series 490; American Chemical Society: Washington, D.C., 1992; pp 74-86.
8. Güntert, M.; Brüning, J.; Emberger, R.; Hopp, R.; Köpsel, M.; Surburg, H.; Werkhoff, P. In *Flavor Precursors. Thermal and Enzymatic Conversions*; Teranishi, R.; Takeoka, G. R.; Güntert, M., Eds.; ACS Symposium Series 490; American Chemical Society: Washington, D.C., 1992; pp 140-163.
9. Tressl, R.; Garbe, L.-A.; Haffner, T.; Lange, H. In *Flavor Analysis: Developments in Isolation and Characterization;* Mussinan, C. J.; Morello, M. J., Eds.; ACS Symposium Series 705, American Chemical Society, Washington, D.C., 1997; pp 167-180.
10. Tressl, R.; Rewicki, D. In *Flavor Chemistry. Thirty Years of Progress;* Teranishi, R.; Wick, E. L.; Hornstein, I. Eds.; Kluwer Academic/Plenum Publishers: New York, 1999; pp 305-325.
11. Takeoka, G. R.; Flath, R. A.; Buttery, R. G.; Winterhalter, P.; Güntert, M.; Ramming, D. W.; Teranishi, R. In *Flavor Precursors. Thermal and Enzymatic Conversions*; Teranishi, R.; Takeoka, G. R.; Güntert, M., Eds.; ACS Symposium Series 490; American Chemical Society: Washington, D.C., 1992; pp 116-138.
12. Herderich, M.; Roscher, R.; Schreier, P. In *Biotechnology for Improved Foods and Flavors*; Takeoka, G.R.; Teranishi, R.; Williams, P.J.; Kobayashi, A., Eds.; ACS Symposium Series 637; American Chemical Society: Washington, D.C., 1996; pp 261-271.
13. Stahl-Biskup, E.; Intert, F.; Holthuijzen, J.; Stengele, M.; Schulz, G. *Flav. Fragr. J.* **1993**, *8*, 61-80.
14. Tominaga, T.; Peyrot des Gachons, C.; Dubourdieu, D. *J. Agric. Food Chem.* **1998**, *46*, 5215-5219.
15. Ohloff, G. *Scents and Fragrances;* Springer Verlag: Berlin, Germany, 1994.
16. Noyori, R. In *Recent Developments in Flavor and Fragrance Chemistry*; Hopp, R.; Mori, K., Eds.; Proceedings of the 3rd International Haarmann & Reimer Symposium; VCH Verlagsgesellschaft mbH: Weinheim, Germany, 1993; pp 3-11.
17. *Enzymatic reactions in organic media;* Koskinen, A. M. P.; Klibanov, A. M., Eds.; Blackie Academic & Professional: Glasgow, UK, 1996.
18. Berger, R.G. *Aroma Biotechnology;* Springer Verlag: Berlin, Germany, 1995.

19. Weber, B.; Maas, B.; Mosandl, A. *J. Agric. Food Chem.* **1995**, *43*, 2438-2441.
20. Schmidt, H.-L.; Rossmann, A.; Werner, R. A. In *Flavorings;* Ziegler, E.; Ziegler, H., Eds.; Wiley-VCH, Weinheim, Germany, 1998.
21. Mosandl, A. *Food Rev. Int.* **1995**, *11*, 597-664.
22. Rothe, R.; Thomas, B. *Z. Lebensm. Unters. Forsch.* **1963**, *109*, 302-310.
23. Guadagni, D. G.; Buttery, R. G.; Okano, S.; Burr, H. K. *Nature (London)* **1963**, *200*, 1288-1289.
24. Grosch, W. *Trends Food Sci. Technol.* **1993**, *4*, 68-73.
25. Milo, C.; Blank, J. In *Flavor Analysis: Developments in Isolation and Characterization;* Mussinan, C. J.; Morello, M. J., Eds.; ACS Symposium Series 705, American Chemical Society, Washington, D.C., 1997; pp 250-259.
26. Guth, H. *J. Agric. Food Chem.* **1997**, *45*, 3027-3032.
27. Guth, H.; Buhr, K.; Fritzler, R. In *Frontiers in Flavour Science;* Schieberle, P.; Engel, K.-H., Eds.; Deutsche Forschungsanstalt für Lebensmittelchemie, Garching, Germany, 2000; pp 235-242.
28. Kubo, I. In *Bioactive Volatile Compounds from Plants*; Teranishi, R.; Buttery, R. G.; Sugisawa, H., Eds.; ACS Symposium Series 525, American Chemical Society, Washington, D.C., 1993; pp 57-70.
29. Vuotto, M. L.; Basile, A.; Moscatiello, V.; De Sole, P.; Castaldo-Cobianchi, R.; Laghi, E.; Ielpo, M. T. L. *Int. J. Antimicrob. Agents.* **2000**, *13*, 197-201.
30. Häusler, A.; Schilling, B. In *Flavour Perception. Aroma Evaluation;* Kruse, H.-P.; Rothe, M., Eds.; Eigenverlag Unversität Potsdam: Bergholz-Rehbrücke, Germany, 1997; pp 375-380.
31. Schwab, W.; Williams, D. C.; Croteau, R. In *Frontiers of Flavour Science*; Schieberle, P.; Engel, K.-H., Eds.; Deutsche Forschungsanstalt für Lebensmittelchemie: Garching, Germany, 2000, pp 445-451.
32. Lancet, D.; Sadovsky, E.; Seidemann, E. *Proc. Natl. Acad. Sci. (USA)* **1993**, *90*, 3715-3719.
33. Reed, R. R. In *Recent Developments in Flavor and Fragrance Chemistry*; Hopp, R.; Mori, K., Eds.; Proceedings of the 3rd International Haarmann & Reimer Symposium; VCH Verlagsgesellschaft mbH: Weinheim, Germany, 1993; pp 275-281.

Instrumental Analysis of Food Flavors

Chapter 2

Odor-Active Compounds of Dry-Cured Meat: Italian-Type Salami and Parma Ham

I. Blank[1], S. Devaud[1], L. B. Fay[1], C. Cerny[2], M. Steiner[2], and B. Zurbriggen[2]

[1]Nestlé Research Center, Nestec Ltd., 1000 Lausanne 26, Switzerland
[2]Nestlé Product Technology Centre, Nestec Ltd., 8310 Kemptthal, Switzerland

The odor-active compounds of Italian-type dry-cured meat products were investigated by gas chromatography in combination with olfactometry and mass spectrometry. A number of key odorants identified in this study have not yet been reported in Italian-type salami and Parma ham. 2-Acetyl-1-pyrroline, methional, 1-octen-3-one, 4-hydroxy-2,5-dimethyl-3(2*H*)-furanone, 4-methylphenol, and sotolone are newly reported in Parma-type ham. 2-Acetyl-1-pyrroline and methional were also unequivocally identified in salami. No single aroma compound eliciting the characteristic salami or Parma ham note could be detected. Most of the odorants found stem from amino acids, lipids, and spices and are generated by fermentation, lipid oxidation or Maillard-type reactions.

The aroma of dry-cured meat products such as salami and ham is significantly different from that of thermally processed meats. While the latter have extensively been studied (review in 1) only limited information is available on key odorants of dry-cured meat products (2). Gas chromatography–olfactometry (GC-O) in combination with dilution techniques has been used for the identification of many sensorially relevant odorants in heated meat (3). GC-sniffing has also been employed as a tool to discriminate odor-active components from odorless volatiles in dry-cured meat products such as ham and sausages (4-7). However, many of the intensely smelling compounds remained unknown due to their low concentrations in fermented meat products (4-6). To the best of our knowledge, GC-sniffing has not yet been applied to Parma-type ham.

In this paper, we present preliminary results obtained on the flavor analysis of Italian-type dry-cured meat products such as salami and Parma-type ham with focus on odor-active compounds using the approach of sensory directed chemical analysis based on GC-O and GC-MS.

Experimental Procedures

Materials

Commercially available samples of Swiss origin were used for flavor analysis. The salami 'La Maza' (Rapelli) was based on pork meat, beef, bacon, salt, sugars, monosodium glutamate, ascorbic acid, spices, and potassium nitrate. The Parma-type ham was made of pork meat and salt. Silica gel 60, anhydrous sodium sulfate, sodium carbonate, diethyl ether (Et_2O), and pentane were from Merck (Darmstadt, Germany). The solvents were freshly distilled on a Vigreux column (1 m x 1.5 cm).

Analysis of Parma Ham Aroma

The ham was homogenized in liquid nitrogen using a Waring Blendor. The ham sample (200 g) was extracted with Et_2O (250 mL) for 1 h. This was repeated twice by adding Et_2O (2 x 200 mL) to the residue and extracting for 2 h each. The organic phase was dried over Na_2SO_4 and concentrated to 100 mL using a Vigreux column (50 x 1 cm). The volatile compounds were separated by distillation in high vacuum at 5 mPa (8). After addition of the solvent extract was completed, the distillation was continued for another 30 min maintaining the temperature at 45 °C. An additional aliquot of Et_2O (50 mL) was added to the residue and the procedure was repeated. The distillates from the two glass traps cooled with liquid nitrogen were collected and concentrated to 1 mL on a Vigreux column and by micro-distillation (9).

GC-O was performed on a Carlo Erba (Mega 2) equipped with a cold on-column injector, flame ionization detector (FID) and a sniffing-port. The effluent was spit 1:1 into a FID and sniffing port (10). A fused silica capillary column of medium polarity (OV-1701) was used, 30 m x 0.32 mm with 0.25 μm film thickness (J&W Sci., Folsom, CA). The temperature program was: 35°C (2 min), 40°C/min to 50°C (1 min), 6°C/min to 180°C, 10°C/min to 240°C (15 min). Linear retention indices were calculated (11).

Electron impact (EI) mass spectra were obtained on a Finnigan MAT 8430 mass spectrometer at 70 eV. Volatile components were introduced *via* a Hewlett-Packard HP-5890 GC using a cold on-column injector. The same type of fused silica capillary column (OV1701) was used as described above. The carrier gas was helium (90 kPa). The temperature program was: 50°C (2 min), 4°C/min to 180°C, 10°C/min to 240°C (10 min).

Analysis of Salami Aroma

Salami without skin (2 kg) was minced, mixed with Na_2SO_4 (2 kg) and homogenized in a Waring Blendor. Portions of 400 g were extracted in a Soxhlet apparatus with pentane/Et_2O (2+1, v/v, 400 mL) during 3 h at 40 °C. The combined solvent extracts (4 L) were concentrated (1.5 L) using a Vigreux column and stored at –20 °C. The extract was distilled under high vacuum (5 mPa) in portions of 250 mL (8) by adding it dropwise into the distillation flask during 1 h. After the addition was completed, the distillation was continued for another hour by maintaining the temperature at 40 °C. The volatile compounds were collected in glass traps cooled with liquid nitrogen. The combined distillates were concentrated to 200 mL on a Vigreux column.

The acidic fraction (A) was obtained by extracting the distillate with aqueous Na_2CO_3 (5 %, 2 x 50 mL). The pH of the aqueous phase was adjusted to 3.0 and re-extracted with Et_2O (2 x 100 mL). Fraction A as well as the remaining organic phase (neutral fraction N) were dried over Na_2SO_4 and concentrated to 25 mL. From fractions A and N, 5 mL each were concentrated to 0.5 mL for GC analysis.

For further fractionation, fraction N (15 mL) was concentrated to 1 mL and separated into 5 subfractions by flash chromatography (12) on silica gel (25 g, pore size 30-60 μm, 20 x 2 cm glass column) using 150 mL of the pentane (N1), pentane/Et_2O (95 + 5, v/v, N2), pentane/Et_2O (80 + 20, v/v, N3), pentane/Et_2O (50 + 50, v/v, N4), and Et_2O (N5). The subfractions were concentrated to 0.2 mL for GC analysis.

GC-O and GC-MS analyses were performed on a HP 5890 GC combined with a HP 5970 MS. The fused silica capillary column employed was an Ultra-1 (SE-30, 50 m x 0.32 mm, 0.52 μm film thickness, Hewlett Packard) with a HP-1 precolumn (5 m x 0.53 mm, 2.65 μm film thickness). At the end of the capillary, the effluent was split 1:1 into a sniffing port and the ion source. The carrier gas was helium (150 kPa). The sample (1 μL) was introduced via the cold on-column technique. The temperature program was: 50°C, 4°C/min to 250°C (30 min). The transfer line temperature was 280 °C. Mass spectra were obtained in the EI mode at 70 eV.

The sensory significance of each odorant was evaluated by Aroma Extract Dilution Analysis (AEDA) and expressed as flavor dilution (FD) factor (3). The FD factors were determined by stepwise dilution of the original aroma extract with Et_2O obtaining FD-factors of 5, 25, and 125.

Results and Discussion

Parma-type Ham

GC analysis of the aroma extract with a concentration factor of 200 (200 g ham → 1 mL extract) resulted in a complex gas chromatogram composed of more than 100 volatile substances. GC-O was employed to screen the odor-active regions.

Fifteen of them showed odor intensities of 1-2 and higher, on a scale from 1 to 3. Particularly intense odor notes were acidic, sweaty, potato-like, fatty, phenolic, seasoning-like, and honey-like. However, no single odorant had a cured, meaty, Parma-type note, as also stated by Piotrowski et al. earlier (13).

Most of the intensely smelling odorants were identified based on retention index and mass spectra or aroma quality (Table I). They showed identical sensory and analytical properties as the reference compounds which were available for structure confirmation. The results summarized in Table I indicate methional (no. 7), (*E*)-2-nonenal (no. 11), and sotolone (no. 14) as the most potent odorants which have not yet been reported in Parma ham. However, (*E*)-2-nonenal was described as volatile constituent of Iberian ham (14).

Further odorants with medium odor intensity were acetic acid (no. 1), isovaleric acid (no. 6), *p*-cresol (no. 13), and phenylacetic acid (no. 15). So far, only acetic and isovaleric acid have been reported in raw Italian-type ham (15). Finally, ethyl isovalerate (no. 2), butyric acid (no. 3), 2-acetyl-1-pyrroline (no. 4), phenylacetaldehyde (no. 9), and 4-hydroxy-2,5-dimethyl-3(2*H*)-furanone (no. 10, HDF) were identified as odor-active compounds. Neither the roasty smelling 2-acetyl-1-pyrroline nor the caramel-like HDF have been reported in Parma ham.

Italian-type Salami

To facilitate identification of minor compounds, the solvent extract of salami was separated into an acidic (A) and a neutral (N) fraction. The major odorants of the fraction A were acidic, butyric and isovaleric acids identified by GC-MS. The gas chromatogram of fraction N, with a concentration factor of 800, was composed of more than 100 volatile compounds (Figure 1).

GC-O resulted in more than 30 odor-active regions showing very different odor qualities. Again, no single odorant could be detected representing the cured, meaty, salami-like note. The predominant odor qualities perceived by the sniffing technique were sulfury (no. 16), green (no. 22), potato-like (no. 7), roasty (no. 4), and nutty (no. 25) with the highest FD-factor of 125.

In addition, fruity (nos. 20, 2), pepper-like (no. 23), mushroom-like (no. 8), musty (nos. 26, 27), and soapy/fatty (nos. 30, 35) odor qualities were found with FD-factors of 25. Identification experiments were focused on these intensely smelling compounds. As shown in Table II, hexanal (no. 22), methional (no. 7), and 2-acetyl-1-pyrroline (no. 4) were identified as the most potent odorants followed by ethyl isobutyrate (no. 20), ethyl isovalerate (no. 2), α-pinene (no. 23), 1-octen-3-one (no. 8), and (*E*)-2-octenal (no. 26).

Table I. Odor-active Compounds Found in the Solvent Extract of a Parma-type Ham.

No.	*Flavor compound*	*Odor quality (GC-O)*	*Retention index*[a]	*Odor intensity (GC-O)*[b]	*Literature*[e]
1	Acetic acid [c]	Acidic, pungent	790	2	15
2	Ethyl isovalerate [d]	Fruity	915	1-2	16
3	Butyric acid [c]	Sweaty	990	1-2	15
4	2-Acetyl-1-pyrroline [d]	Roasty	1015	1-2	- [f]
5	Unknown	Fatty, tallow-like	1020	1-2	
6	Isovaleric acid [c]	Sweaty, musty	1035	2	15
7	Methional [d]	Potato-like	1050	2-3	- [f]
8	1-Octen-3-one [d]	Mushroom-like	1075	1-2	- [f]
9	Phenylacetaldehyde [c]	Honey-like	1185	1-2	17
10	4-Hydroxy-2,5-dimethyl-3(2*H*)-furanone (HDF) [d]	Caramel-like, sweet	1250	1-2	- [f]
11	(*E*)-2-Nonenal [d]	Fatty, leather-like	1275	2-3	14 [g]
12	Unknown	Musty	1285	1-2	
13	*p*-Cresol [c]	Phenolic, musty	1305	2	- [f]
14	Sotolone [d]	Seasoning-like	1355	2-3	- [f]
15	Phenylacetic acid [d]	Honey-like, spicy	1510	2	- [f]

a : The retention indices on OV-1701 are slightly shifted to higher values (as compared to the reference substances) due to the high amount of some short-chain fatty acids.

b : The aroma was described at the sniffing port upon GC-O. The intensity scale was from 1 (weak) to 3 (intense). Only odorants with aroma intensities higher than 1 are listed.

c : Identification is based on comparison of the retention indices (RI) on OV-1701, odor quality, and mass spectra of the reference substance and odorant found in the solvent extract.

d : Identification is based on comparison of the RI data on the capillary column OV-1701 and odor quality with that of the reference compounds. The concentration was too low for unequivocal identification by GC-MS.

e : Reported for the first time in reference 14 (Garcia et al., 1991), 15 (Giolitti et al., 1971), 16 (Barbieri et al., 1992), 17 (Hinrichsen & Pedersen, 1995).

f : Reported for the first time in this study as constituent of Parma-type ham.

g : Reported only in Spanish dry-cured ham.

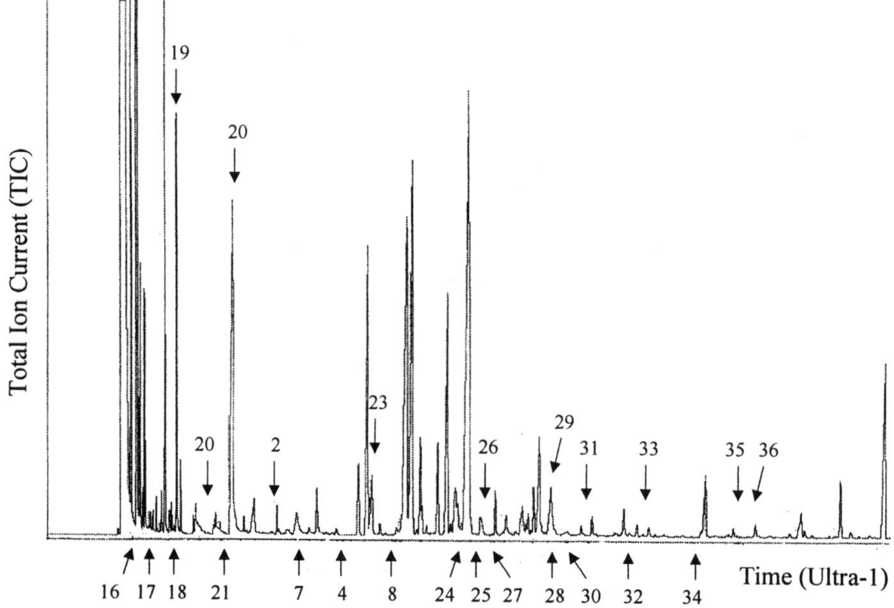

Figure 1. Gas chromatogram of the neutral fraction of Italian-type salami aroma extract obtained on an apolar capillary column. The numbers correspond to those in Table II.

Table II. Odor-active Compounds Found in the Neutral Fraction of an Italian-type Salami.

No.	*Aroma compound*	*Odor quality (GC-O)*	*RI*[a]	*FD-Factor*[b]	*Sub-fraction*	*Literature*[e]
16	Unknown	Sulfury	< 700	125	N2-5	
17	Unknown	Musty, tallowy	< 700	5	N2	
18	Unknown	Buttery, fruity	< 700	5	N2	
19	Allyl methyl sulfide [c]	Garlic -like	< 700	5	N3-5	7
20	Ethyl isobutyrate [d]	Fruity	739	25	N1,2	5
21	Unknown	Fruity, sour	753	5	N2,3	
22	Hexanal [c]	Green	771	125	N2	18
2	Ethyl isovalerate [c]	Fruity, sweet	835	25	N2,3	5
7	Methional [d]	Cooked potato	875	125	N2,3	- [f]
4	2-Acetyl-1-pyrroline [c]	Roasty, popcorn	924	125	N5	19 [g]
23	α-Pinene [c]	Pepper-like	935	25	N1,2	18
8	1-Octen-3-one [d]	Mushroom-like	960	25	N3	5 [g]
24	Unknown	Citrus-like, soapy	1020	5	-	
25	Unknown	Roasty, nutty	1028	125	N3,4	
26	(*E*)-2-Octenal [c]	Musty	1038	25	-	7
27	Unknown	Musty, earthy	1049	25	N3	
28	Nonanal [c]	Citrus, tallowy	1100	5	N2,3	18
29	Linalool [c]	Flowery	1100	5	N3	18
30	Unknown	Citrus-like, soapy	1119	25	N2	
31	Unknown	Fruity, soapy	1136	5	N2,3	
32	Ethyl octanoate [c]	Fatty, green	1177	5	N1	5
33	(*E*,*E*)-2,4-Nonadienal [c]	Fatty	1195	5	N2	- [f]
34	Unknown	Fruity, berry-like	1235	5	-	
35	Unknown	Fatty, meaty	1278	25	N3	
36	(*E*,*E*)-2,4-Decadienal [c]	Fatty, tallowy	1294	5	N4	5

a : Retention indices on Ultra-1.

b : Flavor Dilution (FD) Factor. Only FD-factors higher than 1 are listed.

c : Identification based on comparison of retention indices (RI), odor quality, and mass spectra (MS) of the odorant and reference substance.

d : Tentatively identified (no reference compound available or concentration too low for unequivocal identification by GC-MS).

e : In salami for the first time reported by Schmidt & Berger (7), Berger et al (18), and Stahnke (19). In model sausages identified by Stahnke (5).

f : Reported for the first time in this study as constituent of salami.

g : Tentatively identified.

Most of the odorants listed in Table II have already been reported in the literature. Newly identified odorants were methional (no. 7) and (*E,E*)-2,4-nonadienal (no. 33). The presence of 1-octen-3-one (no. 8) in air-dried model sausages was reported by Stahnke (5). Hexanal (no. 22) was previously found in Swedish fermented sausages (20). As shown in Table II, however, some odor-active compounds were not identified, such as nos. 16, 25, 27, 30, and 35.

2-Acetyl-1-pyrroline was unequivocally identified in the salami sample by GC-MS (Figure 2). Its presence in salami and Parma-type ham was further confirmed by coinjection with the reference sample showing identical GC properties on two capillary columns. During the 9th Weurman Symposium, Stahnke suggested the presence of 2-acetyl-1-pyrroline in Italian and French sausages (19), however without showing supporting analytical data.

The identification of 2-acetyl-1-pyrroline by GC-MS was facilitated by applying isolation techniques with high recovery yields, such as the Solvent Assisted Flavor Extraction (SAFE) recently described by Engel et al. (21). As shown in Table III, distillation under vacuum using SAFE results in higher recovery yields compared to the conventional technique. The model mixture was composed of odorants with a wide range of volatility and polarity dissolved in Et_2O containing 10 % MCT (medium chain triglyceride) to simulate solvent extracts obtained from food.

Formation of Odorants

The chemical structures of the odorants identified in this study are shown in Figure 3. Their formation can be explained by various pathways such as lipid oxidation, fermentation, and Maillard-type reactions. However, some of the odorants might be generated by more than one pathway. Typical lipid oxidation products are 1-octen-3-one (no. 8) and the aldehydes nos. 11, 22, 26, 28, 33, and 36. They are formed by oxygen-mediated radical reactions of unsaturated fatty acids (22). In general, these odorants contribute with fatty notes which should not be too strong to avoid off-flavor formation.

The esters nos. 2, 20, and 32 are most likely formed by microbial esterification of the corresponding short-chain fatty acids with ethanol which are present in fermented meat products. Fatty acids such as butyric acid (no. 3) and isovaleric acid (no. 2) contribute as odorants, but they also function as precursors for fermentative ester production. A further fermentation product is acetic acid. Free fatty acids derive from lipids and (branched) amino acids formed during ripening (23).

Maillard-type reaction products found in this work were Strecker aldehydes such as methional (no. 7) and phenylacetaldehyde (no. 9) and sugar degradation products like HDF (no. 10) and sotolone (no. 14). They belong to the most potent odorants identified in this study. 2-Acetyl-1-pyrroline seems to play an important role in fermented meat products by contributing a strong roasty note. Its formation in bread and heated model systems was shown to be linked to 1-pyrroline, the

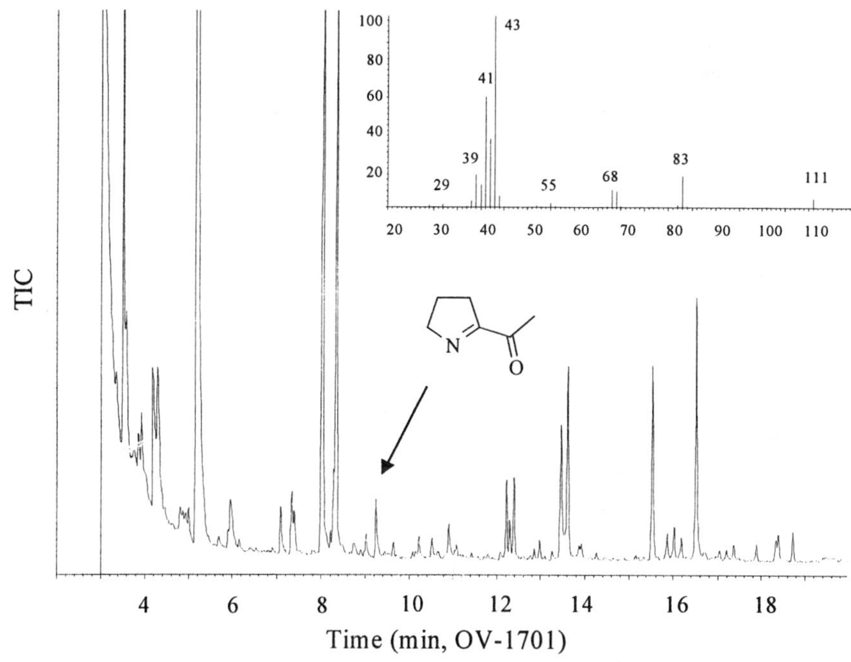

Figure 2. Identification of 2-acetyl-1-pyrroline in an Italian-type salami from the basic fraction of an aroma extract obtained by distillation in vacuum using the SAFE technique (21).

Table III. Recovery yields of selected odorants from a model mixture.

Odorant	*Recovery yields (%) using the technique of*	
	Sen et al. (8)	*Engel et al. (21)*
1-Octen-3-one	40	98
2-Furfurylthiol	35	99
Methyl caprylate	17	87
(*E*,*E*)-2,4-Decadienal	2	34
4-Hydroxy-2,5-dimethyl-3(2*H*)-furanone (HDF)	1	54

Strecker product of proline, and 2-oxopropanal as sugar degradation product (24). It may also be of microbial origin as shown by Romanczyk et al. (25).

Finally, some of the odorants identified stem from spices or feedstuff of plant origin, e.g. allyl methyl sulfide (no. 19), α-pinene (no. 23), and linalool (no. 29). Although sausage flavor can be modulated by microbial combinations (26), the crucial role of spices like pepper and garlic has been emphasized (7).

Conclusions

This study focused on the identification of odor-active compounds and resulted in a number of odorants that contribute to the overall aroma of Italian-Type salami and ham. Some of them have not yet been reported in such fermented meat products, mainly because of the limitations of the analytical tools used. In this work, GC-O has helped to screen and also to identify potent odorants. In general, the aroma quality of a volatile component in combination with retention indices is equivalent to identification by GC-MS provided that the reference compound is available (27). The presence of an odorant can then be verified by coelution with the reference substance on capillaries of different polarity. This approach is helpful for the identification of odorants with very low threshold values and unique aroma qualities.

Catabolism of amino acids, lipids and carbohydrates plays an important role in dry-cured meat flavor formation. Amino acids seem to be essential in flavor formation of dry-cured meat products, e. g. valine, leucine, isoleucine, methionine, proline, and phenylalanine, as a number of potent odorants are structurally related to them. A better understanding of the role of various mechanisms in flavor formation may help to optimize fermentation conditions for the generation of dry-cured aromas.

Acknowledgments

We thank K. Bähler for expert technical assistance and Dr. Elizabeth Prior for linguistic proofreading.

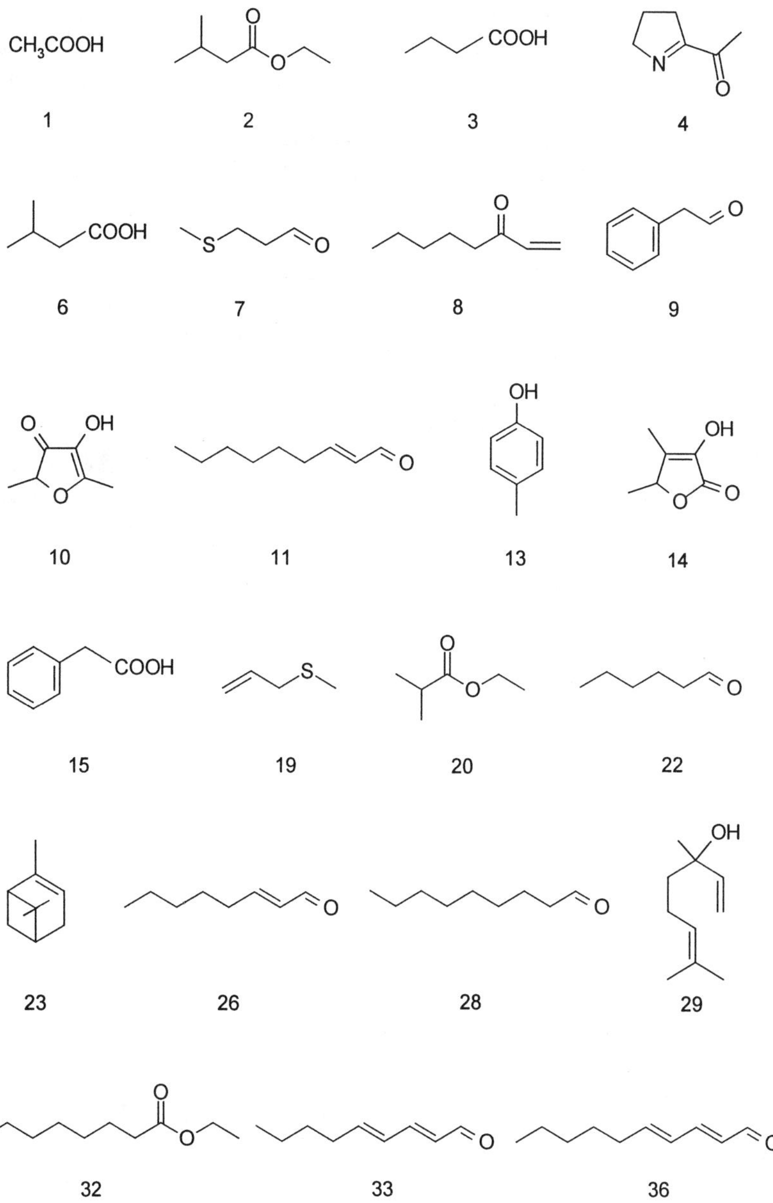

Figure 3. Chemical structures of odorants identified in this study. The numbers correspond to those in Table I and II.

References

1. *Flavor of Meat, Meat Products and Seafoods;* Shahidi, F., Ed.; Blackie Academic Professional: London, UK, 1998, Chapters 2-6, pp 5-130.
2. Flores, M.; Spanier, A. M.; Toldra, F. In *Flavor of Meat, Meat Products and Seafoods;* Shahidi, F., Ed.; Blackie Academic Professional: London, UK, 1998, pp 320.
3. Grosch, W. *Trends Food Sci. Technol.* **1993**, *4*, 68.
4. Berdagué, J. L.; Bonnaud, N.; Rousset, S.; Touraille, C. In *37th Internat. Congr. Meat Sci. Technol.*, 1991, Proceedings Vol. 3, pp 1135.
5. Stahnke, L. H. *Meat Science* **1994,** *38*, 39.
6. Flores, M.; Grimm, C. C.; Toldrá, F.; Spanier, A. M. *J. Agric. Food Chem.* **1997**, *45*, 2178.
7. Schmidt, S.; Berger, R. G. *Lebensm. Wissensch. Technol.* **1998**, *31*, 559.
8. Sen, A.; Laskawy, G.; Schieberle, P.; Grosch, W. *J. Agric. Food Chem.* **1991**, *39*, 757.
9. Bemelmans, J. M. H. In *Progress in Flavour Research;* Land, D. G., Nursten, H. E., Eds.; Applied Sciences: London, UK, 1979, pp 79.
10. Blank, I.; Lin, J.; Devaud, S.; Fumeaux, R.; Fay, L. B. In *Spices – Flavor Chemistry and Antioxidant Properties;* Risch, S. J., Ho, C.-T, Eds.; American Chemical Society: Washington, D.C., 1997, pp 12.
11. van den Dool, H.; Kratz, P. *J. Chromatogr.* **1963**, *11*, 463.
12. Still, W. C.; Kahn, M.; Mitra, J. *J. Org. Chem.* **1978**, *43*, 2923.
13. Piotrowski, E. G.; Zaika, L. L.; Wasserman, A. E. *J. Food Sci.* **1970**, *35*, 321.
14. Garcia, C.; Berdagué, J. J.; Antequera, T.; López-Bote, C.; Córdoba, J. J.; Ventanas, J. *Food Chem.* **1991**, *41*, 23.
15. Giolitti G.; Cantoni, C. A.; Bianchi, M. A.; Renon, P. *J. Appl. Bact.* **1971**, *34*, 51.
16. Barbieri, G.; Bolzoni, L.; Parolari, G.; Virgili, R.; Buttini, R.; Careri, M.; Mangia, A. *J. Agric. Food Chem.* **1992**, *40*, 2389.
17. Hinrichsen, L. L.; Pedersen, S. B. *J. Agric. Food Chem.* **1995**, *43*, 2932.
18. Berger, R. G.; Macku, C.; German, J. B.; Shibamoto, T. *J. Food Sci.* **1990**, *55*, 1239.
19. Stahnke, L. H. In *9th Weurman Flavour Research Symposium*, Freising, Germany, 22-25 June 1999, in press.
20. Halvarson, H. *J. Food Sci.* **1973**, *38*, 310.
21. Engel, W.; Bahr, W.; Schieberle, P. *Eur. Food Res. Technol.* **1999**, *209*, 237.
22. Grosch, W. In *Autoxidation of Unsaturated Lipids;* Chan, H. W. S., Ed.; Academic Press: London, UK, 1987, pp 95.
23. Lücke, F. K. *Food Res. Internat.* **1994**, *27*, 299.
24. Schieberle, P. *J. Agric. Food Chem.* **1995**, *43*, 2442.
25. Romanczyk, L. J. Jr.; McClelland, C. A.; Post, L. S.; Aitken, W. M. *J. Agric. Food Chem.* **1995**, *43*, 469.
26. Berdagué, J. L.; Monteil, P.; Montel, M. C.; Talon, R. *Meat Science* **1993**, *35*, 275.
27. Blank, I. In *Techniques for Analyzing Food Aroma;* Marsili, R., Ed.; M. Dekker: New York, 1997, pp 293.

Chapter 3

Aroma-Active Benzofuran Derivatives: Analysis, Sensory Properties, and Pathways of Formation

Peter Winterhalter and Bernd Bonnländer

Institut für Lebensmitteltechnologie, Technische Universität Braunschweig, Schleinitzstrasse 20, D–38016 Braunschweig, Germany (email: P.Winterhalter@tu-bs.de)

3a,4,5,7a-Tetrahydro-3,6-dimethyl-3*H*-benzofuran-2-one (wine lactone), 2,4,5,7a-tetrahydro-3,6-dimethyl-benzofuran (linden ether), and 2,3,3a,4,5,7a-hexahydro-3,6-dimethyl-benzofuran (dill ether) are typical odorants of white wines, linden honey, and dill herb, respectively. Whereas different chemical syntheses of the benzofuran derivatives have been described and the analysis of stereoisomeric mixtures has been achieved by chiral capillary gas chromatography, pathways of formation of these key flavor compounds are still largely obscure. This paper presents results on wine lactone formation during fermentation and storage of wine and discusses the hydrolytic chemistry of a likely precursor in wine medium. Moreover, first results concerning dill ether formation are presented.

The most important feature of an aroma compound is its sensory contribution to the odor pattern of a complex mixture of volatiles. There are several strategies to determine the flavor contribution of an individual compound. An initial approach uses the ratio of the concentration of a constituent to its odor threshold concentration expressed as so-called 'flavor-units' (*1*). The flavor units show how much the actual concentration of a substance exceeds its threshold concentration. If all of the individual constituents in a flavor mixture together with their flavor thresholds are known, the volatile compounds can be listed according to their odor effectiveness, expressed as number of flavor units. In reality, this approach often causes problems: (i) due to the large number of volatile constituents in natural aromas flavor thresholds are not available for all constituents, and (ii) in many cases unknown constituents are likely to occur in complex natural isolates. Until all flavorants are characterized and quantified, their flavor units cannot be determined. Consequently, alternative methods are required to determine the odor contribution for each component of a mixture. The techniques most widely used today are the so-called 'CharmAnalysis' and the 'Aroma Extract Dilution Analysis' (AEDA), respectively (*2,3*). The principle of both methods is the same. Both use GC-effluent sniffing to detect the odor-active compounds in a mixture. Through serial dilutions of the aroma extract - and the determinations of so-called 'flavor dilution factors' (in the case of AEDA) -, those constituents can be detected which have the highest odor potency. Many successful applications of CharmAnalysis and AEDA have been reported (*4,5*).

Analysis of Aroma-active Benzofuran Derivatives

AEDA of extracts from white wines (*6*) and linden honey (*7*) has led to the identification of potent odorants with a 3,6-dimethyl-benzofuran carbon skeleton. The compounds which were only present in minute amounts have been named wine lactone **1** and linden ether **3**, respectively (cf. Fig. 1). Similarily on the basis of GC-olfactometry, dill ether **2** (*8*) was confirmed as a major aroma contributor of dill seed and dill herb (*9*). Structurally related compounds with a benzofuran skeleton that have been previously isolated from peppermint oil are also shown in Fig. 1. They include mintlactone and isomintlactone (*10*) as well as menthofuran and its oxidation product hydroxy-menthofurolactone (*10-12*). The dehydrated form of the latter compound - the so-called dehydromintlactone - is also known to occur as trace constituent in peppermint oil (*13*). Most recently, a series of novel perhydrobenzofuran derivatives was identified in Italo-Mitcham black peppermint oil (*48*).

All of the above mentioned aroma compounds are chiral, i.e. they contain one or more asymmetric carbon atom(s) and exhibit optical activity. Enantiodifferentiation was of fundamental interest, since the enantiomeric distribution in nature does not only reflect the enantioselectivity of biosynthetic pathways (important for authenticity control), it also determines the character and intensity of the aroma impression. Structure-odor correlations of many naturally occurring chiral flavor compounds have been published, indicating distinct differences in the aroma of two optical antipodes (*14*). For the separation of 3,6-dimethyl-benzofuran derivatives high resolution gas chromatography on modified cyclodextrin phases or on a Chirasil-Val column has been used (*15-21*). Data about suitable phases for chiral-gas chromatography (c-GC) are gathered in Table I together with data about organoleptic properties of naturally-occurring stereoisomers.

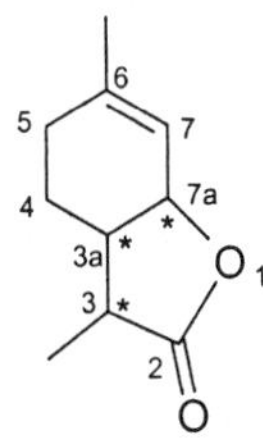

Wine lactone

3a,4,5,7a-Tetrahydro-3,6-dimethyl-3*H*-benzofuran-2-one (**1**)

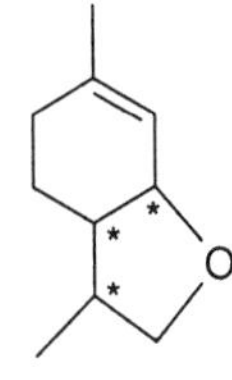

Dill ether

2,3,3a,4,5,7a-Hexahydro-3,6-dimethyl-benzofuran (**2**)

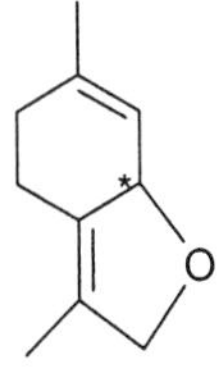

Linden ether

2,4,5,7a-Tetrahydro-3,6-dimethyl-benzofuran (**3**)

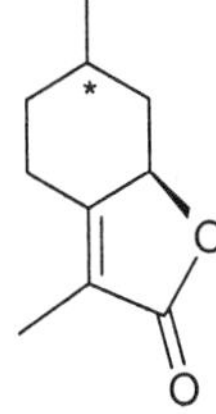

Mintlactone

(7a*R*)-5,6,7,7a-Tetrahydro-3,6-dimethyl-4*H*-benzofuran-2-one

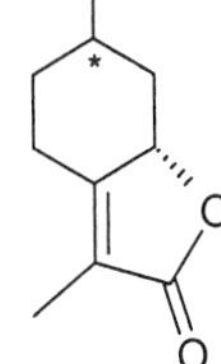

Isomintlactone

(7a*S*)-5,6,7,7a-Tetrahydro-3,6-dimethyl-4*H*-benzofuran-2-one

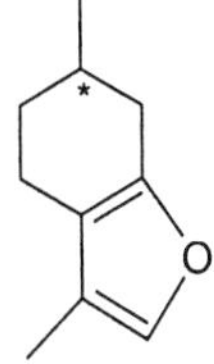

Menthofuran

4,5,6,7-Tetrahydro-3,6-dimethyl-benzofuran

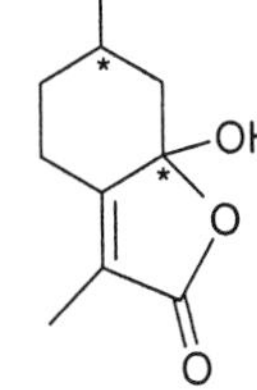

Hydroxymenthofurolactone

5,6,7,7a-Tetrahydro-7a-hydroxy-3,6-dimethyl-4*H*-benzofuran-2-one

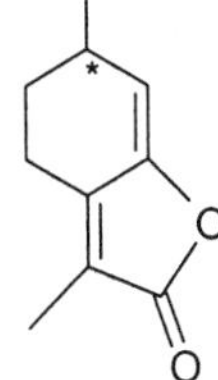

Dehydromintlactone

5,6-Dihydro-3,6-dimethyl-4*H*-benzofuran-2-one

Figure 1. Aroma-active benzofuran derivatives identified in natural sources.

Table I. Occurrence, Separation, and Sensory Properties of Aroma-active Benzofuran Derivatives

Compound	*Natural Isomer (Source)*	*Sensory Properties*	*Flavor Threshold*	*c-GC Separation*	*Ref.*
Wine lactone (8 isomers)	3S,3aS,7aR (wine)	sweet, coconut	0.02 pg/L (air)	3-BUT-2,6-DIPENT-γ-CD	(*15*)
Dill ether (4 stable isomers)	3S,3aS,7aR (dill herb)	typical dill	30 ng/L (air)	6-ME-2,3-DI-PENT-γ-CD Dibutyryl-TBDMS-γ-CD	(*16*) (*17*)
Linden ether (2 isomers)	racemic (linden honey, *Tilia cordata*)	flowery, mint-like	1.5 ng/L (air)	PENT-ß-CD	(*18*)
Mintlactone (2 isomers)	6R,7aR (peppermint oil)	coconut, coumarin	2.8 ng/L (air)	3-TFA-2,6-DI-PENT-ß-CD	(*21*) (*49*)
Isomintlactone (2 isomers)	6R,7aS (peppermint oil)	coconut, coumarin	1.25 ng/L (air)	Chirasil-Val	(*19*) (*49*)
Menthofuran (2 isomers)	6R (peppermint & poley oil) 6S (bucco leaf oil)	minty, herbal, tobacco herbal, turpentine	0.4 ng/L (air)	PME-ß-CD	(*20*) (*49*)
HO-mentho-furolactone (4 isomers)	(+) (bucco leaf oil) *stereochemistry not assigned*	-	-	PME-ß-CD	(*20*)
Dehydromint-lactone (2 isomers)	6R (peppermint oil)	coconut, coumarin	3.1 ng/L (air)	3-TFA-2,6-DI-PENT-ß-CD	(*21*) (*49*)

Abbreviations: 3-BUT-2,6-DIPENT-γ-CD: octakis (3-O-butyryl-2,6-di-O-pentyl)-γ–cyclodextrin; 6-ME-2,3-DIPENT-γ-CD: octakis(6-O-methyl-2,3-O-pentyl)-γ–cyclodextrin; Dibutyryl-TBDMS-γ-CD: octakis (2,3-di-O-butyryl-6-O-*tert*-butyldimethylsilyl)-γ-cyclodextrin; PENT-ß-CD: heptakis (2,3,6-tri-O-pentyl)-ß-cyclodextrin; PME-ß-CD: heptakis-(2,3,6-tri-O-methyl)-ß-cyclodextrin; 3-TFA-2,6-DI-PENT-ß-CD: heptakis (3-O-trifluoroacetyl-2,6-di-O-pentyl)-ß-cyclodextrin.

'Wine lactone' - 3a,4,5,7a-Tetrahydro-3,6-dimethyl-3*H*-benzofuran-2-one (1)

Lactone **1** was first isolated by Southwell (*22*) as essential oil metabolite from the urine of koala bears. Later it has been obtained in the course of chemical syntheses (*23,24*) but without recognizing its sensory potential. Only in recent years in the course of GC-olfactometric studies, Guth (*6*) discovered lactone **1** as one of the most important flavor compounds in wine. Hence, the lactone was named 'wine lactone' **1**. Synthesis of all eight stereoisomers has been reported by Guth (*15*) together with data on their odor thresholds and sensory properties. Importantly, the naturally occurring (3*S*,3a*S*,7a*R*)-stereoisomer of wine lactone was reported to possess an extraordinarily low flavor threshold of 0.00002 ng/L of air.

In the case of Gewürztraminer wine, changes in concentration of wine lactone **1** after yeast and malolactic fermentation, respectively, as well as after maturation have been monitored by Guth (*25*). Whereas yeast fermentation gave only a slight increase in the concentration of **1** (from 0.03 to 0.06 µg/L), subsequent storage (4 months) in high-grade steel tanks resulted in another threefold increase (from 0.06 to 0.19 µg/L). This latter observation clearly indicates that the generation of **1** in wine medium is rather a chemical than a biochemical process.

From 100 L of a German Riesling wine, the glucose ester of (*E*)-2,6-dimethyl-6-hydroxy-2,7-dienoic acid (**4a,** cf. Fig. 2) could be isolated and characterized (*26,27*). In addition to the glucoconjugated form **4a**, free (*E*)-2,6-dimethyl-6-hydroxy-2,7-dienoic acid (**4**) was also present in Riesling wine. The structure of the aglycone and the entire glucoconjugate was confirmed by synthesis (*28*). In order to study the hydrolytic chemistry of acid **4**, model degradation reactions (SDE, pH 3.2, 2.5, and 2.0) have been carried out (*28*). Importantly, in all cases wine lactone **1** was detected as a major conversion product of acid **4**. Additional degradation products are also outlined in Fig. 2. They include the dehydration products **7** and **8a/b** as well as the tentatively identified isomeric monoterpenoid acids **9a/b**. Mass spectral data for the degradation products have been published previously (29).

In order to study the influence of the fermentation process on wine lactone formation from precursor **4,** model fermentations with oenological yeast as well as different strains of 'wild' yeasts have been carried out. In each case, 30 mg of acid **4** were added to the medium (minimal and full media, initial pH was 6, for details cf. ref. *28*) during the growth phase of the yeast. The fermentation broth was liquid-liquid extracted with Et_2O and the level of wine lactone **1** was analyzed by GC-MS. The data are presented in Table II. In all cases, only trace amounts of wine lactone **1** could be detected. Since in the control experiment (acid **4** in culture medium without yeast) the target compound **1** was present at almost the same level as in the fermentation experiments, yeast fermentation has no influence on wine lactone formation. The small amounts of lactone **1** formed during fermentation are obviously the result of a hydrolytic breakdown of acid **4** even in weak acidic medium.

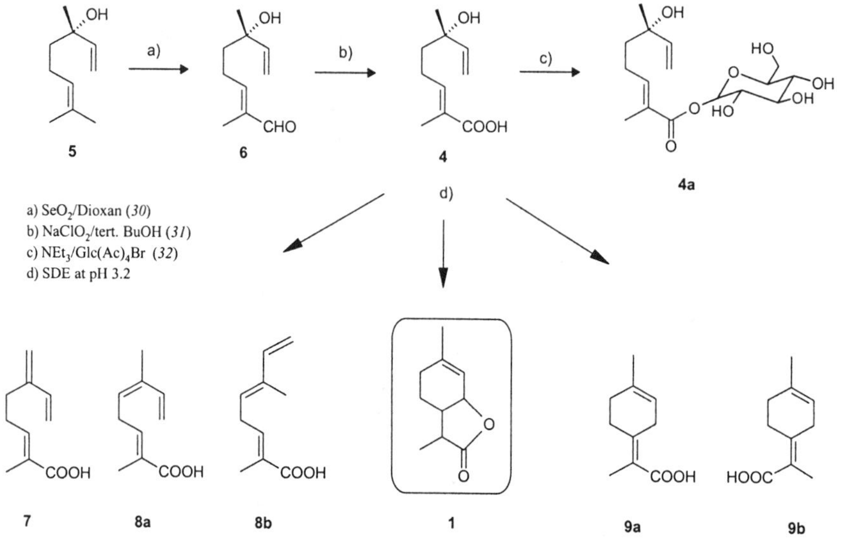

OH
a)
b)
c)
CHO
COOH
HO
OH
5
6
4
4a
d)
a) SeO_2/Dioxan (30)
b) $NaClO_2$/tert. BuOH (31)
c) NEt_3/Glc(Ac)$_4$Br (32)
d) SDE at pH 3.2
7
8a
8b
1
9a
9b
HOOC

*Figure 2. Syntheses of free and glucosidically bound (E)-2,6-dimethyl-6-hydroxy-2,7-dienoic acid (**4/4a**) from linalool (**5**) and the hydrolytic conversion (SDE, pH 3.2) of **4** into lactone **1** as well as degradation products **7-9** (28).*

Table II. Influence of Yeast Fermentation on Wine Lactone Formation

Yeast strain / CBS[1] No.	*Concentration[2] (experiment)*	*Concentration[2] (blank)*	*Δ[3]*
Saccharomyces cerevisiae 6303	0.37	0.44	- 0.07
Saccharomyces cerevisiae 4054	0.25	0.28	- 0.03
Saccharomyces bayanus 3008	0.31	0.35	- 0.04
Saccharomyces bayanus 1177	0.35	0.45	- 0.10
Candida stellata 843	0.36	0.44	- 0.08
Torulaspora delbrueckii 728	0.26	0.31	- 0.05
Torulaspora delbrueckii 3085	0.30	0.28	+ 0.02
Zygosaccharomyces bailii 4691	0.27	0.31	- 0.04

Notes: [1]Centraalbureau voor Schimmelcultures, Baarn, NL; [2]Concentration of lactone **1** relative to internal standard (1-phenylethanol); [3]difference in rel. concentrations of **1** in experiment - blank.

As to the formation of wine lactone during maturation of wine, storage experiments have been carried out. 5 mg of acid **4** have been stored in model wine medium (water/ethanol, 9:1; pH 3.2) at 40°C. After 12, 24 and 30 months, HRGC and HRGC-MS analyses of hydrolytically formed wine lactone **1** have been carried out. HRGC-MS analysis revealed a similar pattern of volatile degradation products as obtained under the more drastic reaction conditions of atmospheric SDE (cf. Fig. 2). Wine lactone **1** indicated by a single peak on a DB-5 column was again detected as the major degradation product. Although the target compound was only formed in a low yield (ratio of wine lactone **1** to acid **4** was approximately 2 % after 30 months), the overall degradation rate of acid **4** did continuously increase with time. This observation is likely to explain the reported increase in wine lactone concentration during storage of wine. Due to the extremely low levels of lactone **1** in wine, conclusive c-GC separations of the natural product have been unsuccessful so far.

Since acid **4** is a common metabolite of the monoterpene alcohol linalool, wine lactone is expected to occur in many additional natural sources. So far, lactone **1** was *inter alia* detected in orange juice (*50*), basil leaves (*33*) and black pepper (*34*). Although the formation of wine lactone **1** in acidic medium through hydrolytic breakdown of acid **4** has been established, its biogenesis in grapes and other natural products still remains to be elucidated.

'Dill ether' - 2,3,3a,4,5,7a-Hexahydro-3,6-dimethyl-benzofuran (2)

Benzofuran **2** was isolated from dill herb oil in 1977 by Bélafi-Réthy and Kerényi (*8*). Due to its low flavor threshold and the typical dill-note it is considered as character impact compound of fresh dill herb (*9,35-39*). The content of ether **2** in dill herb increases during the growth period and reaches its maximum at the flowering stage with a relative amount of ca. 30 % of total dill oil (*40*). Although a straightforward synthesis of dill ether had been published as early as 1966 by Ohloff *et al.* (*41*), it took until the year 1984 to determine its absolute configuration. Applying Ohloff's synthetic approach and using optically pure limonene as starting material, Brunke and Rojahn (*42*) were able to synthesize four stable cis-fused isomers of dill ether **2** and assigned (3*S*,3a*S*,7a*R*)-configuration to the naturally-occurring stereoisomer. A first c-GC separation of dill ether and its enantiomer was reported by König *et al.* (*16*). A novel enantioselective synthesis of dill ether isomers and their separation by c-GC has recently been published by Reichert *et al.* (*17*). With regard to the biogenesis of **2**, GC-IRMS measurements led to the assumption that α-phellandrene and dill ether **2** might be biogenetically closely related (*43*). More details concerning dill ether formation are not known since labeling and feeding experiments have not been carried out so far. On the other hand, different chemical pathways are known that lead to the odoriferous ether **2**. In the course of model degradation studies carried out with 8-hydroxy-linalool (**10**) (cf. Fig. 3), Strauss and coworkers (*44*) observed a distinct dill-note in the hydrolysate. The authors were able to detect trace amounts of dill ether **2**, *p*-menth-1-en-9-al **15**, as well as the tentatively identified trienols **13** and **14a/b** (Fig. 3). As major conversion product of diol **10**, the allylic rearrangement product **12** was formed. When we repeated the experiment of Strauss et al., two additional conversion products of diol **10** could be detected, i.e. 8-hydroxy-nerol **11** and uroterpenol **16** (*45*).

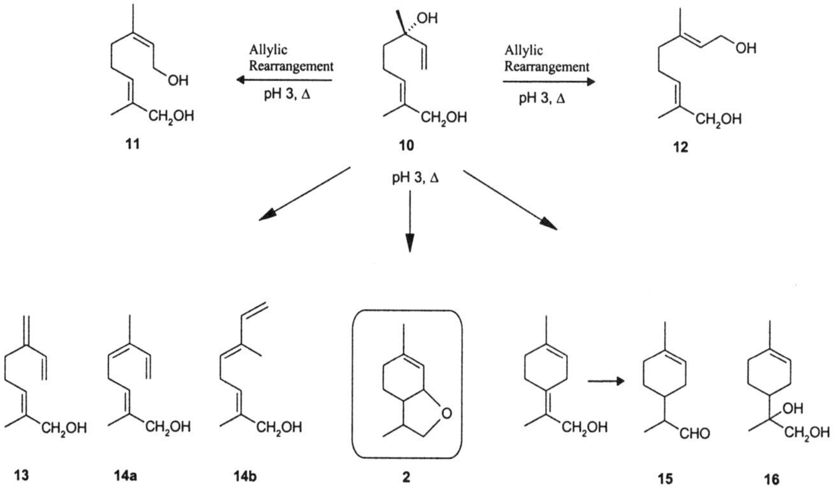

*Figure 3. Acid-catalyzed degradation of linalool derivative **10** (44,45).*

Another known progenitor of dill ether **2** is *p*-menth-2-en-1,9-diol **18** (cf. Fig. 4) which had been used by Ohloff and coworkers in its first synthesis (*41*). Diol **18** can be prepared by photooxidation of *p*-menth-1-en-9-ol **17**. Due to the formation of other photooxidation products, the overall yield of ether **2** from **17** is less than 20 %. A more recent synthesis of dill ether uses methyl ester **19** (*24*). Through reduction to *p*-menth-1-en-3,9-diol **20** and subsequent cyclization, dill ether **2** is obtained in almost 80 % yield.

In the course of our studies on aroma progenitors in dill herb, we were most recently able to isolate substantial amounts (60 mg from 10 kg herb) of the monoterpene glucoside **21** (cf. Fig. 4). 9-Hydroxy-piperitone ß-D-glucopyranoside **21** has been identified for the first time in nature and its structure has been unambiguously elaborated by one and two dimensional NMR studies (*46*). Determination of the absolute configuration is in progress. Although feeding experiments have not yet been performed, glucoside **21** can be considered as one of the missing links in dill ether biosynthesis. Through stereoselective reduction of the keto-function and subsequent enzyme-mediated cyclization the formation of optically-pure dill ether **2** can be rationalized. Experiments to prove this hypothesis are subjects of our actual research.

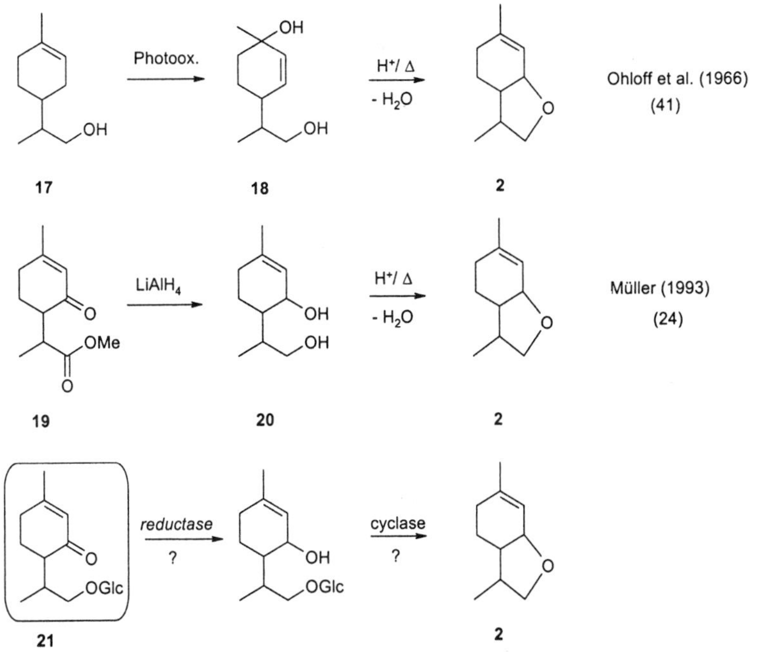

Photoox.
H^+/Δ
$- H_2O$
Ohloff et al. (1966) (41)
17 **18** **2**

$LiAlH_4$
H^+/Δ
$- H_2O$
Müller (1993) (24)
19 **20** **2**

reductase ?
cyclase ?
21 **2**

*Figure 4. Known synthetic routes to dill ether **2** and the postulated generation from the recently isolated 9-hydroxy-piperitone ß-D-glucoside **21** (46).*

'Linden ether' - 2,4,5,7a-Tetrahydro-3,6-dimethyl-benzofuran (3)

With regard to the remaining benzofuran derivatives outlined in Fig. 1, we have only started to investigate the formation of linden ether **3**. Ether **3** was found to occur as a racemate in blossoms of the lime tree (*Tilia cordata*) as well as in linden honey (*7,18*). Only recently an enantioselective synthesis of **3** has been published (47). The detection of racemic mixtures of volatiles always raises questions about their genuineness. Also in the case of linden ether **3**, a secondary formation from a reactive precursor is considered as being likely. One candidate is the reactive allyldiol **22** which has been postulated as progenitor of the odoriferous ethers **3a/b** (cf. Fig. 5) already ten years ago (*18*). Even under mild acidic conditions diol **22** is expected to cyclize to the target compounds **3a/b**. The non-volatile aroma precursors of 10 kg of linden blossoms have been isolated by XAD-2 column chromatography. Aliquots of the methanolic eluate have been subjected to SDE treatment at pH 2.5. GC-MS analysis of the volatiles formed under acidic conditions revealed a formation of dill ether **3a/b**. This observation clearly indicates the presence of a non-volatile progenitor in the XAD-2 isolates. The isolation of the genuine precursor of linden ether **3** is in progress.

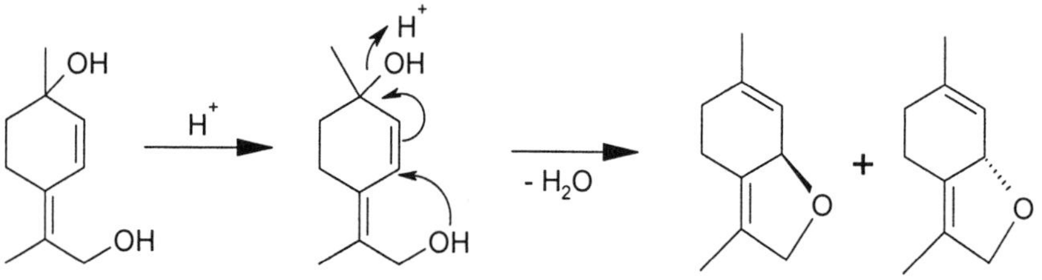

*Figure 5. Postulated formation of racemic linden ether **3a/b** from p-mentha-2,4(8)-dien-1,9-diol (22).*

Acknowledgments

We thank Prof. R.G. Berger and F. Neuser (Universität Hannover) for the fermentation studies and Dr. G. Krammer (H&R, Holzminden) for GC-FTIR analyses. The skillful assistance of A. Bode, U. Harnischmacher, and B. Hübner is gratefully acknowledged.

References

1. Rothe, M.; Thomas, B. *Z. Lebensm. Unters. Forsch.* **1963**, *119*, 302-310.
2. Acree, T.E.; Barnard, J.; Cunningham, D.G. *Food Chem.* **1984**, *14*, 273-286.
3. Schmid, W.; Grosch, W. *Z. Lebensm. Unters. Forsch.* **1986**, *182*, 407-412.

4. Acree, T.E. In *Flavor Science - Sensible Principles and Techniques*; Acree, T.E.; Teranishi, R., Eds.; American Chemical Society: Washington, DC, 1993, pp. 1-20.
5. Grosch, W. *Chem. unserer Zeit* **1990**, *24*, 82-89.
6. Guth, H. *J. Agric. Food Chem.* **1997**, *45*, 3022-3026.
7. Blank, I.; Fischer, K.-H., Grosch, W. *Z. Lebensm. Unters. Forsch.* **1989**, *189*, 426-433.
8. Bélafi-Réthy K.; Kerényi, E. *Acta Chim. Acad. Scient. Hungaricae* **1977**, *94*, 1-9.
9. Blank, I.; Grosch, W. *J. Food Sci.* **1991**, *56*, 63-67.
10. Takahashi, K.; Someya, T.; Muraki, S.; Yoshida, T. *Agric. Biol. Chem.* **1980**, *44*, 1535-1544.
11. Treibs, W. *Ber. Dtsch. Chem. Ges.* **1937**, *70*, 85-88.
12. Wienhaus, H. *Angew. Chem.* **1934**, *47*, 415.
13. Hopp, R. In *Recent Developments in Flavor and Fragrance Chemistry*; Hopp, R.; Mori, K. (Eds.); VCH: Weinheim, Germany, 1993, pp. 13-28.
14. Koppenhoefer, B.; Behnisch, R.; Epperlein, U.; Holzschuh, H.; Bernreuther, A.; Piras, P.; Roussel, C. *Perfumer & Flavorist* **1994**, *19*, 1-14.
15. Guth, H. *Helv. Chim. Acta* **1996**, *79*, 1559-1571.
16. König, W.A.; Icheln, D.; Runge, T.; Pforr, I.; Krebs, A. *J. High Resolut. Chromatogr.* **1990**, *13*, 702-707.
17. Reichert, S.; Wüst, M.; Beck, T.; Mosandl, A. *J. High Resolut. Chromatogr.* **1998**, *21*, 185-188.
18. Blank, I.; Grosch, W.; Eisenreich, W.; Bacher, A.; Firl, J. *Helv. Chim. Acta* **1990**, *73*, 1250-1257.
19. Chavan, S.P.; Zubaidha, P.K.; Ayyangar, N.R. *Tetrahedron Lett.* **1992**, *33*, 4605-4608.
20. Werkhoff, P.; Brennecke, S.; Bretschneider, W.; Schreiber, K. *H&R Contact* **1995**, *64*, 7-11.
21. Werkhoff, P.; Brennecke, S.; Bretschneider, W.; Güntert, M.; Hopp, R.; Surburg, H. *Z. Lebensm. Unters. Forsch.* **1993**, *196*, 307-328.
22. Southwell, I.A. *Tetrahedron Lett.* **1975**, *24*, 1885-1888.
23. Bartlett, P.A.; Pizzo, C.F. *J. Org. Chem.* **1981**, *46*, 3896-3900.
24. Müller, E. *Thesis*, FU Berlin, Germany, 1993.
25. Guth, H. In *Chemistry of Wine Flavor*; Waterhouse, A.L.; Ebeler, S.E. (Eds.); ACS Symp. Ser. 714; American Chemical Society: Washington, DC, 1998, pp. 39-52.
26. Winterhalter, P.; Messerer, M.; Bonnländer, B. *Vitis* **1997**, *36*, 55-56.
27. Bonnländer, B.; Baderschneider, B.; Messerer, M.; Winterhalter, P. *J. Agric. Food Chem.* **1998**, *46*, 1474-1478.
28. Bonnländer, B.; Cuevas-Montilla, E.; Winterhalter, P. *Vitis* (submitted).

29. Winterhalter, P.; Baderschneider, B.; Bonnländer, B. In *Chemistry of Wine Flavor*; Waterhouse, A.L.; Ebeler, S.E., Eds.; ACS Symp. Ser. 714; American Chemical Society: Washington, DC, 1998, pp. 1-12.
30. Hirata, T.; Aoki, T.; Hirano, Y.; Ito, T. Suga, T. *Bull. Chem. Soc. Jpn.* **1981**, *54*, 3527-3529.
31. Sekine, T.; Fukasawa, N.; Ikegami, F.; Saito, K.; Fujii, Y.; Murakoshi, I. *Chem. Pharm. Bull.* **1997**, *45*, 148-151.
32. Lehmann, H.; Schütte, R. *Journal f. prakt. Chemie* **1977**, *319*, 117-122.
33. Guth, H.; Murgoci, A.-M. In *Flavour Perception - Aroma Evaluation*, Proceedings of the 5th Wartburg Aroma Symposium; Kruse, H.-P.; Rothe, M. (Eds.); Eigenverlag Universität Potsdam: Bergholz-Rehbruecke, Germany, 1997, pp. 233-242 .
34. Jagella, T.; Grosch, W. *Eur. Food Res. Technol.* **1999**, *209*, 16-21.
35. Schreier, P., Drawert, F.; Heindze, I. *Lebensm.-Wiss. u.-Technol.* **1981**, *14*, 150-152.
36. Huopalahti, R. *Lebensm.-Wiss. u.-Technol.* **1986**, *19*, 27-30.
37. Brunke, E.J.; Hammerschmidt, F.J.; Koester, F.H.; Mair, P. *J. Essent. Oil Res.* **1991**, *3*, 257-267.
38. Blank, I.; Sen, A.; Grosch, W. *Food Chem.* **1992**, *43*, 337-343.
39. Charles, D.J.; Simon, J.E.; Widrlechner, M.P. *J. Essent. Oil Res.* **1995**, *7*, 11-20.
40. Huopalahti, R.; Linko, R.R. *J. Agric. Food Chem.* **1983**, *31*, 331-333.
41. Ohloff, G.; Schulte-Elte, K.H.; Willhalm, B. *Helv. Chim. Acta* **1966**, *49*, 2135-2150.
42. Brunke, E.J.; Rojahn, W. *Dragoco Report* **1984**, *3*, 67-74.
43. Faber, B.; Bangert, K.; Mosandl, A. *Flavour Fragr. J.* **1997**, *12*, 305-314.
44. Strauss, C.R.; Wilson, B.; Williams, P.J. *J. Agric. Food Chem.* **1988**, *36*, 569-573.
45. Bonnländer, B. *Thesis*, Technische Universität Braunschweig, Germany (in preparation).
46. Bonnländer, B. Winterhalter, P. submitted to *J. Agric. Food Chem.*
47. Brocksom, U.; Toloi, A.P.; Brocksom, T.J. *J. Braz. Chem. Soc.* **1996**, *7*, 237-242.
48. Näf, R.; Velluz, A. *Flavour Fragr. J.* **1998**, *13*, 203-208.
49. Güntert, M.; Krammer, G.; Lambrecht, S.; Sommer, H.; Surburg, H.; Werkhoff, P. In *Aroma Active Compounds in Foods: Chemistry and Sensory Properties*; Takeoka, G.; Engel, K.-H.; Güntert, M., Eds.; ACS Symp. Ser.; American Chemical Society: Washington, DC, 2000.
50. Hinterholzer, A.; Schieberle, P. *Flavour Fragr. J.* **1998**, *13*, 49-55.

Chapter 4

Volatile Constituents of Asafoetida

Gary Takeoka

Western Regional Research Center, Agricultural Research Service, U.S. Department of Agriculture, 800 Buchanan Street, Albany, CA 94710

Asafoetida is the oleogum resin obtained by incision of the roots of various plants from the genus *Ferula* (family Umbelliferae) indigenous to Central Asia, Afghanistan and Iran. Asafoetida has a strong, tenacious, sulfurous odor and is an important spice used in Indian, European and some Middle Eastern foods. Since recent studies reported that asafoetida extracts possessed antifungal activity we investigated the chemical composition of this spice. Volatiles were isolated from *Ferula asafoetida* L. by dynamic headspace sampling and characterized using GC and GC-MS. This paper reports the composition of volatiles and discusses the flavor contribution of individual volatiles.

Asafoetida is the oleogum resin obtained from various plants from the genus *Ferula* (family Umbelliferae) such as *F. alliacea* Boiss., *F. asafoetida* L., and *F. foetida* Regel, that are grown mainly in Afghanistan, Iran, Pakistan and to some extent India. There are about sixty species of *Ferula* with the most important commerical species described by Raghavan et al. (*1*) according to their geographical occurrence and usage. They grow from perennial root stocks and reach a height of four to ten feet. The gum resin is obtained as a latex by incision of the living roots. Asafoetida is distinguished as asafoetida hing (hing) and asafoetida hingra (hingra). Hing is derived chiefly from *F. alliacea* and also from *F. asafoetida* L. This material is used for flavoring purposes whereas hingra, obtained from *F. foetida*, is used for medicinal purposes. Due to its strong flavor characteristics Hing is diluted before being marketed. It is typically diluted with gum arabic, rice flour, corn flour or wheat flour and sold in this compounded form. Asafoetida is used in traditional Chinese medicine where it is viewed as entering the liver, spleen and stomach channels. It stimulates the intestinal, respiratory and the nervous system and is used treat food stagnation, weak digestion, intestinal parasites and flatulence.

Asafoetida has a bitter acrid taste and a characteristic odor similar to that of onion but stronger and more tenacious. It is an important condiment in India and Iran, used to flavor foods such as curries, meatballs and pickles (*2*). It has been used in perfumes and for flavoring in Europe and the United States. Asafoetida absolute has been described to possess an immensely rich and sweet-balsamic body of highly interesting type beneath the garlic-onion like topnote (*3*). It has been demonstrated that an extract of *Ferula narthex* Boiss. did not contain any sulfur constituents (*4*).

The major constituents of asafoetida are the resin (40-64%), gum (25%) and essential oil (10-17%) (*5*). A later study reported a similar range of oil content (6.7-19.6%) (*1*). Three sesquiterpene coumarin ethers, farnesiferol A, B and C, containing monocyclic or bicyclic terpenoid moieties (Fig. 1), were identified in the non-volatile fraction of asafoetida (*6,7*). In contrast, studies by Appendino and co-workers (*8*) revealed that the coumarinic fraction derived from their samples was made up mostly of acyclic oxygenated farnesol derivatives. These researchers also revised the structure of the previously reported asacoumarin B and revealed that this compound is actually galbanic acid (Fig. 1).

It has been known for some time that sulfur compounds are responsible for the characteristic odor of asafoetida. Knowledge of sulfur compounds in asafoetida dates back to the late 1890s when Semmler (*9*) first identified a disulfide with a boiling point of 83-84°C/9mmHg. Mannich and Fresenius (*10*) confirmed that this disulfide which constituted 40% of the oil was 2-butyl propenyl disulfide. Reduction of this major disulfide with zinc yielded levorotatory 2-butanethiol. Asafoetida oil is dominated by three sulfur compounds, 2-butyl (*E,Z*)-1-propenyl disulfide, 1-(methylthio)propyl (*E,Z*)-1-propenyl disulfide and 2-butyl 3-(methylthio)-2-propenyl disulfide (Fig. 2) (*11,12*). In their study of the composition of different asafoetida oils, Abraham *et al.* (*12*) reported that these three compounds constituted 80 to 90% of the oils with the proportion of 2-butyl (*E,Z*)-1-propenyl disulfide, 1-(methylthio)propyl (*E,Z*)-1-propenyl disulfide and 2-butyl 3-(methylthio)-2-propenyl disulfide ranging from 36-84%, 9-31% and 0-52%, respectively. The latter compound was not found in three of the four asafoetida samples from Afghanistan (*12*). Kjaer and co-workers (*13*) separated the *E*- and *Z*-isomers of 2-butyl 1-propenyl disulfide by pressure liquid chromatography and determined the *E,Z*-ratio to be 70:30 while Noleau *et al.* (*14*) found a similar ratio (67:33) by capillary gas chromatography. It has been determined that the natural *E*- and *Z*-isomers of this levorotatory disulfide occur predominantly in the (R)-configuration with an enantiomeric purity of about 75% (*13*) though previous studies *(9,10,11)* suggest that this purity may vary considerably depending on origin of the gum resin. Dimethyl trisulfide, 2-butyl methyl disulfide and di-2-butyl disulfide were previously identified by Rajanikanth *et al.* (*15*). A comprehensive study of asafoetida volatiles was conducted by Noleau *et al.* (*14*) who identified 82 constituents in samples from Pakistan and Iran.

Asafoetida has been reported to have an inhibitory effect on both seed colonization and aflatoxin production by *Aspergillus flavus* (NRRL-3000) (*16*). Additionally, the ethanol extract of this spice has been found to inhibit the growth of four common food spoilage fungi (*17*). Our study was conducted to elucidate the active principle(s) responsible for this antifungal action.

Experimental

Materials. Powdered asafoetida (ingredients: rice flour, asafoetida (*Ferula assa-foetida* L.), gum arabic) was obtained from Frontier (Norway, IA).

Sample Preparation. Dynamic Headspace Sampling. The powdered asafoetida sample (80 g) was placed in a 2 L round-bottomed flask along with 400 mL purified

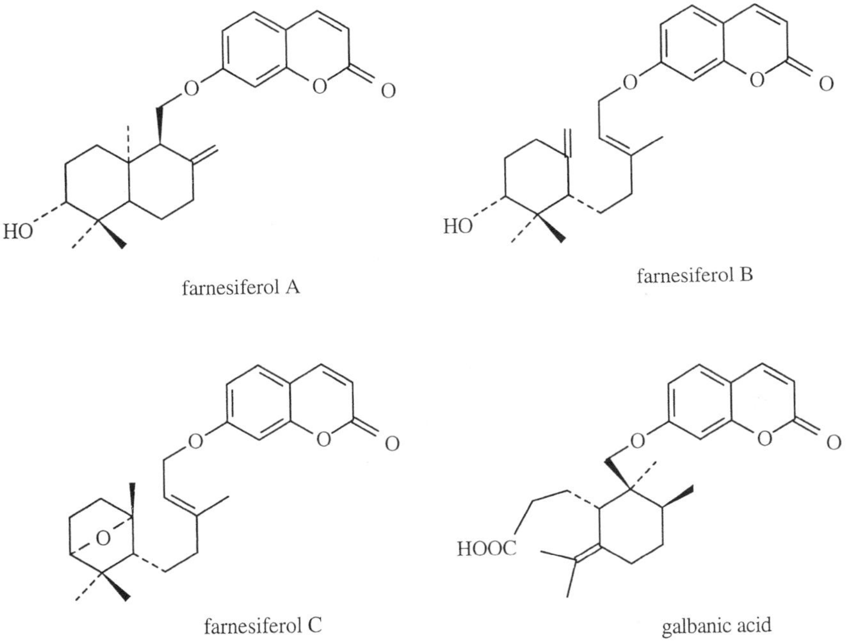

HO

farnesiferol A

HO

farnesiferol B

farnesiferol C

HOOC

galbanic acid

Figure 1. Structures of farnesiferol A, B and C and galbanic acid.

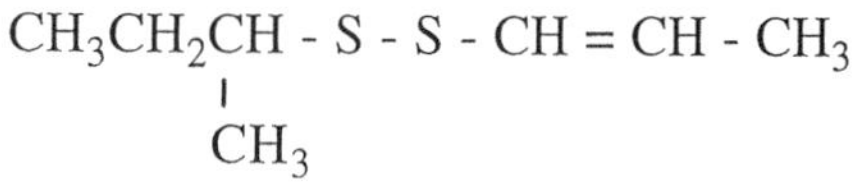

$$CH_3CH_2\underset{\displaystyle CH_3}{\underset{|}{CH}} - S - S - CH = CH - CH_3$$

2-butyl 1-propenyl disulfide

$$CH_3CH_2\underset{\displaystyle SCH_3}{\underset{|}{CH}} - S - S - CH = CH - CH_3$$

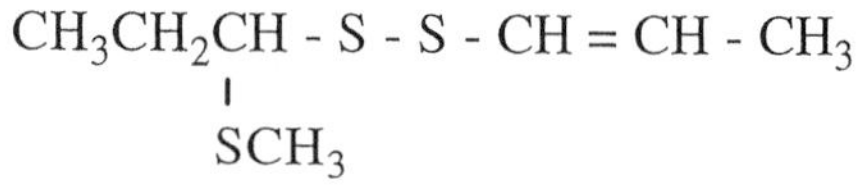

1-(methylthio)propyl 1-propenyl disulfide

$$CH_3CH_2\underset{\displaystyle CH_3}{\underset{|}{CH}} - S - S - CH_2 - CH = \underset{\displaystyle SCH_3}{\underset{|}{CH}}$$

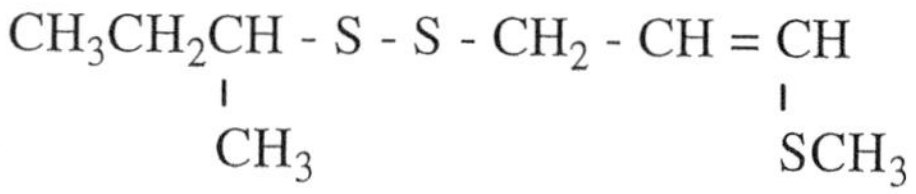

2-butyl 3-(methylthio)-2-propenyl disulfide

Figure 2. Three major sulfur compounds identified in asafoetida.

water (Milli-Q Plus, Millipore Corporation, Bedford, MA) and 216 g NaCl (previously heated to 150 °C to remove volatiles). The flask was fitted with a Pyrex head to allow the sweep gas to enter the top of the flask (via a Teflon tube) and exit out of a side arm through a Tenax trap (ca. 10 g of Tenax [Alltech Associates, Deerfield, IL], fitted with ball and socket joints). The system was purged with nitrogen (200-400 ml/min) for 2 min and immediately connected to an all Teflon diaphram pump that recirculated nitrogen around the loop (closed loop sampling) for at 6 L/min for 3 h. The sample was continuously stirred during the sampling period with a magnetic stirrer. After the sampling the Tenax trap was removed and the volatiles eluted with 50 ml of freshly distilled diethyl ether containing *ca.* 0.001% Ethyl antioxidant 330 (1,3,5-trimethyl-2,4,6-tris(3,5-di-tert-butyl-4-hydroxybenzyl)benzene). The ether was carefully concentrated to ca. 50 μL using a warm water bath (50-60 °C) and a Vigreux column.

Capillary Gas Chromatography. A Hewlett-Packard 6890 gas chromatograph equipped with a flame ionization detector (FID) was used. A 60 m X 0.32 mm i.d. DB-1 fused silica capillary column (d_f = 0.25 μm), J&W Scientific, Inc., Folsom, CA) was employed. The injector and detector temperatures were 180 °C and 290 °C, respectively. The oven temperature was programmed from 30 °C (4 min isothermal) to 200 °C (held for 25 min at final temperature) at 2 °C/min. The helium carrier gas linear velocity was 36 cm/s (30 °C). A Hewlett-Packard 5890 gas chromatograph equipped with an FID and a 60 m X 0.32 mm i.d. DB-WAX fused silica capillary column (d_f = 0.25 μm) was also used. The injector and detector temperatures were 180 °C and 290 °C, respectively. The oven temperature was programmed from 30 °C (4 min isothermal) to 170 °C (held for 25 min at final temperature) at 2 °C/min. The helium carrier gas linear velocity was 35 cm/s (30 °C).

Capillary Gas Chromatography-Mass Spectrometry (GC/MS). Two systems were employed. The first system consisted of a HP 6890 gas chromatograph coupled to a HP 5973 quadrupole mass spectrometer (capillary direct interface). A 60 m X 0.25 mm i.d. DB-1 fused silica capillary column (d_f = 0.25 μm) was used. Helium carrier gas was used at a column headpressure of 16 psi. The oven temperature was programmed from 30 °C (4 min isothermal) to 200 °C at 2 °C/min. The second system consisted of a HP 5890 gas chromatograph coupled to a HP 5971 quadrupole mass spectrometer (capillary direct interface). A 60 m X 0.25 mm i.d. DB-WAX fused silica capillary column (d_f = 0.25 μm) was used. Helium carrier gas was used at a headpressure of 10 psi. The oven temperature was programmed from 30 °C (4 min isothermal) to 175 °C (held for 25 min at final temperature) at 2 °C/min.

Results and Discussion

Volatile constituents of asafoetida were isolated by dynamic headspace sampling of the compounded spice. Sample constituents were identified by comparison of the compound's Kovats index, I (*18*) and mass spectrum with that of a reference standard.

Asafoetida contains a variety of mono- and sesquiterpenoids. The most abundant monoterpenoid was β-pinene (4.341%). With an odor threshold of 6 ppb it is probably a significant contributor to the odor. Other abundant monoterpene hydrocarbons were

Table I. Volatile Constituents of Asafoetida: Headspace Sampling

Constituent	I^{DB-1} exp.	I^{DB-1} ref.	% area[a]
(2-methyl-2-propanethiol)[b,c]			0.040
(2,3-dimethylthiirane)[b,c]			0.044
propyl acetate[c]	699	695	0.038
1-methylthio-(*Z*)-1-propene	706	716[d]	0.028
1-methylthio-(*E*)-1-propene	719	726[d]	0.064
dimethyl disulfide	724	724	0.113
toluene[c]	754	748	0.013
pentanol[c]	757	758	0.196
(S-methyl propanethioate)[b,c]	776		0.071
hexanal	778	778	0.247
(2-(methylthio)butane)[b,c]	797	800	0.197
2-methyl-2-pentenal[c]	808	808	0.107
(3-methyl-2-hexanone)[b,c]	836		0.026
(*E*)-2-hexen-1-ol[c]	856	856	0.070
hexanol[c]	859	848	0.242
2-heptanone[c]	871	865	0.039
3,4-dimethylthiophene	883	898[d]	0.043
methyl (*Z*)-1-propenyl disulfide	906	915[d]	0.092
(2-methyl-3,5-hexadien-2-ol)[b]	915		
methyl (*E*)-1-propenyl disulfide	915	922[d]	0.309
2-ethyl-1-pentanol[c]	922		0.034
α-thujene	923	922	0.043
benzaldehyde[c]	924	926	0.097

α-pinene	929	929	2.888
dimethyl trisulfide	938	941	0.010
camphene	940	941	0.061
(*E*)-2-hepten-1-ol[c]	956	948	0.047
6-methyl-5-hepten-2-one[c]	968	961	
sabinene	968	964	
ß-pinene	968	968	4.341
(2-butyl methyl disulfide)[b,c]	973		0.054
myrcene	984	981	0.284
(2,3,4-trimethylthiophene)[b]	999		0.058
(2-isopropylfuran)[b,c]	1004		0.449
p-cymene	1007	1010	0.089
1,8-cineole[c]		1018	
limonene	1022	1020	1.696
cis-ocimene	1030	1026	0.269
(2-butyl vinyl disulfide)[b,c]	1034		0.066
trans-ocimene	1041	1037	0.552
γ-terpinene	1049	1048	0.040
trans sabinene hydrate	1051	1051	0.039
(2-butyl ethyl disulfide)[b,c]	1054		0.038
trans linalool oxide THF[c]	1057	1056	0.034
cis linalool oxide THF[c]	1070	1070	0.029
2-phenylethyl alcohol[c]	1076	1081	0.124
linalool	1079	1083	0.072
dipropyl disulfide	1085	1085	0.226
pinocarveol	1115	1121	0.073
trans-verbenol[c]	1124	1126	0.095
pinocarvone[c]	1131	1136	0.190
(2-butyl propyl disulfide)[b,c]	1153		

Continued on next page.

Table I. Volatile Constituents of Asafoetida: Headspace Sampling (Continued)

	I^{DB-1}		
Constituent	*exp.*	*ref.*	*% area*[a]
2-butyl-1-propenyl disulfide (*cis*?)	1153		20.488[e]
2-butyl-1-propenyl disulfide (*trans*?)	1158		17.445[e]
terpinene-4-ol[c]	1160	1159	0.298
α-terpineol	1169	1170	0.241
4-vinylphenol[c]	1185	1190	0.079
di-2-butyl disulfide	1194	1194	0.321
(methyl 1-(methylthio)ethyl disulfide)[b,c]	1201		0.033
fenchyl acetate	1204	1205	0.083
4-methoxybenzaldehyde[c] + sub. phenol	1207	1211	0.151
(thymyl methyl ether or carvacrol methyl ether)[b]	1213		0.084
methyl 1-(methylthio)propyl disulfide[c]	1223	1235[f]	0.084
sulfur compound - MW176?	1232		0.083
sulfur compound - MW176?	1243		0.050
4-ethylguaiacol	1248	1248	0.046
1-methoxy-4-(1-propenyl)benzene[c] [anethole]	1256	1256	0.039
bornyl acetate	1265	1268	0.236
4-vinylguaiacol	1278	1280	0.233
sulfur compound - MW182?	1306		0.073
sulfur compound - MW180?	1313		0.328
eugenol	1323	1327	0.089
neryl acetate[c]	1343	1342	0.110
α-longipinene[c]	1350	1350	0.023
sulfur compound - MW178?	1356		0.019

sulfur compound - MW196?	1362		0.328
methyl eugenol	1369	1370	0.047
α-copaene	1374	1374	0.042
(1-(methylthio)propyl propyl disulfide)[b,c]	1404	1397[f]	
1-(methylthio)propyl-1-propenyl disulfide (*cis*?)	1404		31.687[e]
1-(methylthio)propyl-1-propenyl disulfide (*trans*?)	1406		5.097[e]
sulfur compound - MW210?	1443		0.562
(methyl isoeugenol)[b,c]	1459		0.230
ß-selinene	1478	1480	0.094
myristicin[c]	1480	1482	0.086
valencene[c]	1482	1487	0.064
α-selinene[c]	1485	1489	0.071
(1,2,2-trimethyl-1-(p-tolyl)-cyclopentane)[b,c]	1487		0.037
ß-bisabolene	1496	1500	0.138
(delta-cadinol)[b,c]	1502		0.065
(7-epi-alpha-selinene)[b,c]	1508		
δ-cadinene	1508	1514	0.097
trans-γ-bisabolene	1518	1523	0.104
(epi-ligulyl oxide)[b,c]	1538		0.257
(guaiol)[b]	1576		0.025
10-epi-gamma eudesmol	1595		0.046
sulfur compound - MW228?	1611		0.046
2-pentadecanone[c]	1678	1679	0.034
sulfur compound - MW242?	1690		0.174
(farnesyl acetate)[b,c]	1766		0.032
2-hexadecanone[c]	1780	1779	0.016
(2-heptadecanone)[b,c]	1881		0.010

[a]Peak area percentage of total FID area excluding the solvent peaks (assuming all response factors of 1). [b]tentative or partial identifications enclosed in parentheses. [c]identified for the first time in asafoetida. [d]Kuo and Ho (*29*). [e]Peak areas not accurate since isomers were poorly resolved on the DB-1 column. [f]Kuo et al. (*31*).

α-pinene (2.888%), limonene (1.696%), (*E*)-ocimene (0.552%), (*Z*)-ocimene (0.269%) and myrcene (0.284%). α–Pinene also has an odor threshold of 6 ppb and must be an important odor contributor. Limonene is a relatively weak odorant; (R)-(+)-limonene has an odor threshold of 200 ppb while (S)-(-)-limonene has an odor threshold of 500 ppb (*19*). (*Z*)-Ocimene has an odor threshold of 55 ppb while myrcene has an odor threshold of 13 ppb. Smaller amounts of p-cymene (0.089%; odor threshold - 150 ppb), camphene (0.061%), α-thujene (0.043%) and γ-terpinene (0.040%; odor threshold - 65 ppb) were found. Other monoterpenoids identified included terpinen-4-ol (0.298%; odor threshold - 340 ppb), α-terpineol (0.241%; odor threshold - 330 ppb), bornyl acetate (0.236%; odor threshold - 75 ppb), pinocarvone (0.190%), verbenol (0.095%), fenchyl acetate (0.083%), pinocarveol (0.073%), linalool (0.072%; odor threshold - 6 ppb), 1,8-cineole (0.060%; odor threshold - 1.3 ppb) and sabinene hydrate (0.039%).

Three phenylpropanoids, (*E*)-anethole (0.039%), eugenol (0.110%; odor threshold - 11 ppb) and methyl eugenol (0.047%; odor threshold - 68 ppb) were identified. Other phenolics identified included 4-vinylphenol (0.079%; odor threshold - 10 ppb), 4-vinyl-2-methoxyphenol [(4-vinylguaiacol) (0.233%; odor threshold - 3 ppb)] and 4-ethylguaiacol [(4-ethyl-2-methoxyphenol) (0.046%)]. 4-Vinyl-2-methoxyphenol has been found to be an important contributor to popcorn aroma (*20,21*). This compound has been shown to be the main product formed from the thermally induced decarboxylation of ferulic acid (*22*).

Asafoetida oil contains three major sulfur compounds, 2-butyl (*E,Z*)-1-propenyl disulfide, 1-(methylthio)propyl (*E,Z*)-1-propenyl disulfide and 2-butyl 3-(methylthio)-2-propenyl disulfide. The latter compound is only found in certain types of asafoetida and we did not detect this compound in our samples that originated from Pakistan. 2-Butyl 1-propenyl disulfide constituted about 35% of the total volatiles whereas 1-(methylthio)propyl 1-propenyl disulfide accounted for about 36% of the total volatiles. 2-Butyl 1-propenyl disulfide existed in a *E*:*Z* ratio of 51:49 while earlier studies had found a larger proportion of the (*E*) isomer. For 1-(methylthio)propyl 1-propenyl disulfide the *E*:*Z* ratio was 72:28. The isomers of both compounds were much better resolved on the DB-WAX column than on the DB-1 column. The DB-WAX column was used to determine the isomeric ratio. Methyl (*Z*)-1-propenyl sulfide and methyl (*E*)-1-propenyl sulfide were previously identified in steam-distilled onion oil by Boelens and co-workers (*23*). Methyl (*Z*)-1-propenyl disulfide and methyl (*E*)-1-propenyl disulfide, identified in distilled onion oil by Brodnitz et al. (*24*), are reported to possess the odor of cooked onions with an odor threshold of 6.3 μg/L water (*23*). It has been observed that exposure of methyl 1-propenyl disulfide to heat or UV radiation results in the formation of 3,4-dimethylthiophene as the major product (*23*). A small amount of 3,4-dimethylthiophene (0.04%) was identified in our sample. This constituent has an odor threshold of 1.3 μg/L water (*23*). Carson and Wong (*25*) identified dimethyl disulfide in onion distillate. Buttery and co-workers (*26*) found that dimethyl trisulfide is one of the most important flavor compounds in cabbage, broccoli and cauliflower. This compound is a rather potent odorant with an odor threshold of 0.01 ppb. Dipropyl disulfide was identified in the low temperature distillate of fresh onions by Niegisch and Stahl (*27*). Our odor threshold value of 2 nL/L water is in good agreement with the reported value of 3.2 μg/L water (*23*) for this compound.

Methyl 1-(methylthio)propyl disulfide (MeSCHEtSSMe) has been found in the headspace of roasted meat (*28*), in the steam distillate of Welsh onions and scallions (*29*), and in a model reaction system containing propanal, hydrogen sulfide and methanethiol (*30*).

A number of sulfur compounds remain uncharacterized and their mass spectra are listed in Table II.

Table II. Unidentified sulfur compounds and their mass spectra

Compound	*KI*[a]	*%Area*	*Mass Spectrum*
unknown 1	1232	0.084	178(4), 177(4), 176(40), 122(5), 121(3), 120(40), 87(9), 83(100), 59(12), 57(16), 55(100), 53(11), 41(14)
unknown 2	1313	0.328	182(0.7), 181(0.5), 180(5), 107(2), 105(1), 79(4), 77(5), 76(4), 75(100), 59(11), 47(8), 45(10), 41(9)
unknown 3	1362	0.327	210(2), 196(3), 182(3), 181(9), 77(5), 76(4), 75(100), 59(8), 47(5), 41(10)
unknown 4	1443	0.562	210(1), 91(5), 90(5), 89(100), 79(3), 73(6), 61(10), 57(4), 41(20)
unknown 5	1690	0.174	244(0.3), 243(0.2), 242(2), 91(5), 90(5), 89(100), 79(5), 73(8), 61(10), 45(8), 41(21)

[a]Kovats index (DB-1)

The mass spectrum of unknown 4 is similar to that of an unidentified constituent [compound no. 198; 89(100), 61(17), 41(43), 45(14), 39(9), 47(7), 57(6), 91(5), 90(5)] found in leek and asafoetida by Noleau *et al.* (*14*).

A group of compounds whose mass spectra suggest are oxygen containing sesquiterpenes also remain uncharacterized. The large quantity of mono- and sesquiterpenoids as well as sulfur-containing constituents suggest the possibility of sulfur-containing mono- and sesquiterpenoids, but we have not yet identified these constituents in asafoetida.

Topics for future study include the determination of odor thresholds of the major sulfur compounds such as 2-butyl 1-propenyl disulfide and 1-(methylthio)propyl 1-propenyl disulfide. It would also be useful to perform aroma extraction dilution analyses (AEDA) or CHARM analyses on the extracts.

Literature Cited

1. Raghavan, B.; Abraham, K.O.; Shankaranarayana, M.L.; Sastry, L.V.L.; Natarajan, C.P. *Flavour Ind.* **1974**, *5*, 179-181.
2. The New Encyclopaedia Britannica, 15th Edition, Encyclopaedia Britannica, Inc.: Chicago, 1998

3. Arctander, S. In *Perfume and Flavor Materials of Natural Origin*, Elizabeth, N.J., 1960, 76-77.
4. Samini, M.N.; Unger, W. *Planta Med.* **1979**, *36*, 128-133.
5. Guenther, E. In *The Essential Oils*, Vol. 4, D. Van Nostrand Company, Inc., New York, **1960**, p. 570-572.
6. Caglioti, L.; Naef, H.; Arigoni, D.; Jeger, O. *Helv. Chim. Acta* **1958**, *41*, 2278.
7. Caglioti, L.; Naef, H.; Arigoni, D.; Jeger, O. *Helv. Chim. Acta* **1959**, *42*, 2557-2570.
8. Appendino, G.; Tagliapietra, S.; Nano, G.M.; Jakupovic, J. *Phytochemistry* **1994**, *35*, 183-186.
9. Semmler, F.W. *Arch. Pharm.* (*Weinheim*) **1891**, *229*, 1.
10. Mannich, C.; Fresenius, Ph. *Arch Pharm.* (*Weinheim*) **1936**, *274*, 461-462; *Chem. Abstr. 31*, 1951 (1937).
11. Naimie, H.; Samek, Z.; Dolejs, L.; Rehacek, Z. *Collect. Czech. Chem. Commun.* **1972**, *37*, 1166-1177.
12. Abraham, K.O.; Shankaranarayana, M.L.; Raghavan, B.; Natarajan, C.P. *Indian Food Packer* **1979**, *33*, 29-32.
13. Kjaer, A.; Sponholtz, M.; Abraham, K.O. Shankaranarayana, M.L.; Raghavan, R.; Natarajan, C.P. *Acta Chemica Scand.* B **1976**, *30*, 137-140.
14. Noleau, I.; Richard, H.; Peyroux, A.S. *J. Essen. Oil Res.* **1991**, *3*, 241-256.
15. Rajanikanth, B.; Ravindranath, B.; Shankaranarayana, M.L. *Phytochemistry* **1994**, *23*, 899-900.
16. Ghewande, M.P.; Nagaraj, G. *Mycotoxin Research* **1987**, *3*, 19-24.
17. Thyagaraja, N.; Hosono, A. *Lebensm.-Wiss. u.-Technol.* **1996**, *29*, 286-288.
18. Kovats, E., sz. *Helv. Chim. Acta* **1958**, *41*, 1915-1932.
19. Buttery, R.G. In *Flavor Chemistry: Thirty Years of Progress*; Teranishi, R.; Wick, E.L.; Hornstein, I, Eds.; Kluwer Academic/Plenum Publishers: New York, 1999; pp. 353-366.
20. Schieberle, P. *J. Agric. Food Chem.* **1991**, *39*, 1141-1144.
21. Buttery, R.G.; Ling, L.C.; Stern, D.J. *J. Agric. Food Chem.* **1997**, *45*, 837-843.
22. Fiddler, W.; Parker, W.E.; Wassermann, A.E.; Doerr, R.C. *J. Agric. Food Chem.* **1967**, *15*, 757-761.
23. Boelens, M.; de Valois, P.J.; Wobben, H.J.; van der Gen, A. *J. Agric. Food Chem.* **1971**, *19*, 984-991.
24. Brodnitz, M.H.; Pollock, C.L.; Vallon, P.P. *J. Agric. Food Chem.* **1969**, *17*, 760-763.
25. Carson, J.F.; Wong, F.F. *J. Agric. Food Chem.* **1961**, *9*, 140-143.
26. Buttery, R.G.; Guadagni, D.G.; Ling, L.C.; Seifert, R.M.; Lipton, W. *J. Agric. Food Chem.* **1976**, *24*, 829-832.
27. Niegisch, W.D.; Stahl, W.H. *Food Res.* **1956**, *21*, 657.
28. Dubs, P.; Stüssi, R. *Helv. Chim. Acta* **1978**, *61*, 2351-2359.
29. Kuo, M.-C.; Ho, C.-T. *J. Agric. Food Chem.* **1992**, *40*, 111-117.
30. Boelens, M.; van der Linde, L.M.; de Valois, P.J.; van Dort, H.M.; Takken, H.K. *J. Agric. Food Chem.* **1974**, *22*, 1071-1076.
31. Kuo, M.-C.; Chien, M.; Ho, C.-T. *J. Agric. Food Chem.* **1990**, *38*, 1378-1381.

Chapter 5

Gas Chromatographic Analysis of Chiral Aroma Compounds in Wine Using Modified Cyclodextrin Stationary Phases and Solid Phase Microextraction

Susan E. Ebeler[1], Gay M. Sun[2], Allen K. Vickers[3], and Phil Stremple[3]

[1]Department of Viticulture and Enology, University of California, One Shields Avenue, Davis, CA 95616
[2]Aeroject Fine Chemicals, Hazel Avenue and Highway 50, Rancho Cordova, CA 95670
[3]J&W Scientific, Inc., 91 Blue Ravine Road, Folsom, CA 95630–4714

Knowledge of enantiomer composition is key to understanding many aspects of flavor chemistry. As reviewed by Marchelli et al. (*1*) information on sensory properties, source ("natural" *vs*. synthetic), geographic origin, processing/aging treatments, and formation mechanisms (e.g., chemical *vs*. enzymatic reactions) of flavors in foods and beverages can often be assessed based on enantiomeric composition.

Separation of chiral flavors

Beginning in the mid 1800's with Pasteur's separation of tartaric acid enantiomers in wine, chemists have looked for improved methods for the analysis and separation of chiral compounds. Derivatization of analytes with chiral reagents to form diastereomers which can be chromatographically separated on achiral phases has been widely employed. For example, one of the earliest successful GC separations was demonstrated by Casanova and Corey in 1961 (*2;* reviewed by *3*) who separated racemic camphor following derivatization with (R,R)-2,3-butanediol. However, these procedures require use of an enantiomerically pure derivatization reagent and the reaction conditions can often result in racemization of the analyte.

Direct separation of enantiomeric compounds on chiral amide-based GC stationary phases (e.g., Chirasil-val) was demonstrated in the mid 1960's (*4*; reviewed by *3, 5*). These columns allowed for excellent separation of amino acids but had limited application for analysis of most volatile compounds. With the introduction of chiral cyclodextrin-based GC stationary phases in the late 1970's and 1980's, chemists were provided with the ability to directly separate a large number of underivatized chiral compounds. For example, several early studies demonstrated separations of cyclic and acyclic enantiomers with a range of functional groups including lactones, terpene hydrocarbons, carbonyls, alcohols, spiroketals, and oxiranes (*6 - 8*).

Chiral cylodextrin stationary phases for GC have been particularly useful for analysis of flavors in foods and beverages. Takeoka et al. (*9*) described enantiomeric separations of the monoterpene hydrocarbon fraction from ginger oil; sesquiterpene hydrocarbon enantiomers (e.g., α-copaene and δ-elemene) from hops, orange oil, and ginger oil; and the isomers of methyl 2-methylbutanoate from pineapple essence. Separation of chiral sulfur volatiles in passion fruits (*10*) has been accomplished using cyclodextrin stationary phases and is further discussed in other chapters of this proceedings. The effect of processing on the enantiomeric distribution of three important terpenes, limonene, linalool, and α-terpineol, in cherries has been evaluated (*11*). These examples do not provide a comprehensive review of applications of cyclodextrin phases for food and beverage analysis but demonstrate the wide range of compounds that can be separated.

For the analysis of grapes and wines, cyclodextrin stationary phases have been used to establish the enantiomeric distribution of isomeric 3,4-dihydro-3-oxoedulans in Riesling wine; these compounds can contribute a camphoraceous note or a weak tobacco note depending on the structure (*12*). Similarly, determination of enantiomeric distribution of solerone (5-oxo-4-hexanolide) in brandy and Riesling acetal in wine has been accomplished (*13, 14*). Guth (*15*) separated the eight possible isomers of wine lactone (3a, 4, 5, 7a-tetrahydro-3,6-dimethylbenzofuran-2(3H)-one) on a cyclodextrin stationary phase and determined that the predominate isomer occurring in Gewürztraminer wine was the 3S, 3aS, 7aR-isomer which has an intense sweet coconut-like aroma and an aroma threshold of 0.02 pg/L air. Again, while not a comprehensive list of applications, these examples demonstrate the range of compounds that can be separated.

Cyclodextrins are composed of 6, 7, or 8 α-D-glucose units linked together to form a cyclic oligosaccharide having a chiral torus shape. The cyclodextrins can form inclusion complexes with small molecules and the shape and size of the guest molecules can affect the enantioselectivity (*16*). The interior of the cyclodextrin ring is hydrophobic in nature and van der Waals interactions probably dominate separation of chiral hydrocarbons or compounds with large nonpolar regions which can fit into the torus. However, hydrogen bonding and dipole-dipole interactions at the periphery of the ring can also play a role in the interaction and therefore even large molecules which do not fit into the cavity may be enantioseparated (*16*). For gas chromatographic applications, the hydroxyl groups on the glucose molecules are commonly derivatized (methyl, pentyl, acetyl, trifluoroactyl, and t-butyldimethylsilyl derivatives are common) to improve chromatographic properties. In addition, the modified cyclodextrins are dissolved in conventional stationary phases so that low operating temperatures can be used. However, modifications to the cyclodextrin hydroxyl groups will affect both the size of the cavity as well as surface interactions, and interaction mechanisms for these phases have not been widely studied.

Sample Preparation and Solid Phase Microextraction

Sample preparation and extraction of volatile components are critical steps in any flavor analysis. It is during these steps that trace volatiles are concentrated, impurities are removed, or specific fractions are isolated. The composition of the resulting sample can be influenced by the researchers' decisions regarding such variables as the selectivity of the extracting matrix, the potential for artifact formation during isolation, and the time and cost associated with sample preparation. Among the methods that have been commonly used for the analysis of volatile flavors are liquid-liquid extraction, static and dynamic headspace analysis, and distillation/extraction. The advantages and disadvantages of each have been reviewed (*17 - 20*).

Solid Phase MicroExtraction (SPME) offers a rapid, solvent-free method for extracting organic compounds from aqueous samples. Introduced in the late 1980's, SPME was originally applied to the analysis of hydrocarbon contaminants in water. The method has since been used in a variety of fields, including flavor chemistry (*21 - 25*). Fused-silica fibers are coated with a polymeric adsorbent that specifically selects for the target analytes (*26*). These adsorbent-coated fibers are inserted into either the liquid phase or headspace of a sample. Once the fiber has equilibrated, it is removed from the sample and immediately inserted into the inlet of the gas chromatograph where the concentrated analytes are rapidly desorbed in the heated inlet environment.

A number of fiber coatings are now available which can give different selectivity for volatiles analysis, however, in the analysis of alcoholic beverages, adsorption of ethanol to these fibers can limit their application. The large ethanol peak obtained in the chromatograms can interfere with detection of early eluting peaks; use of selective detectors (e.g., Nitrogen-Phosphorous, Flame Photometric) or mass spectrometry can eliminate this interference by using selected ion monitoring or by setting the mass cut-off range above that of ethanol.

Several applications of the use of SPME for analysis of volatiles in alcoholic beverages have recently been published. Evans et al. (*27*) used headspace SPME in combination with gas chromatography-mass spectrometry in the selected ion monitoring (GC-MS/SIM) mode for the accurate quantitation of 2,4,6-trichloroanisole (TCA) in wine. Using a fully deuterated [$^{2}H_{5}$] TCA analog as the internal standard, a limit of quantitation of 5 ng/L was obtained, comparable to existing solvent extraction methods. Similar results were published by Fischer and Fischer (*28*).

Accurate quantitation of diacetyl (2,3-butanedione) in wine is difficult due to its high volatility and relatively low concentrations, and due to interferences from other closely related wine components (*29*). Using headspace SPME with a deuterated internal standard ([$^{2}H_{6}$] diacetyl) and GC-MS/SIM, a detection limit of 0.01 μg/mL with linearity to 10 μg/mL was obtained (*29*). The method was rapid (10 min extraction time) and the diacetyl peak was free of interference from other GC eluents, in contrast to other common analysis procedures (steam distillation and/or derivatization).

De la Calle Garcia et al. (*24*) used headspace SPME for the analysis of terpene volatiles in Riesling wines. They reported results similar to those obtained by liquid/liquid extraction. They also noted that static headspace analysis was an ineffective method for extracting wine aroma compounds of medium volatility, such as the terpenes.

Levels of dimethyl sulfide (DMS), which contributes a cooked corn aroma to beer, are carefully monitored in the brewing industry. Using SPME, the limit of detection for DMS in beer was 1 μg/L, well below the reported sensory threshold of 30-45 μg/L (*30*). The SPME method provided quantitative results that were comparable to a static headspace procedure for DMS in beer but required no prior decarbonation of the beer

In summary, SPME is gaining rapid acceptance as a valuable tool for the analysis of volatile flavors. SPME can be significantly faster and easier than solvent extraction methods; SPME extraction times are commonly less than 60 min compared to several hours for many solvent extraction methods. SPME is easily automated and it does not require the use of potentially toxic and expensive solvents. The development of new phases with differing polarities offers promise for the analysis of a wide range of volatiles, however, further applications to wine and other alcoholic beverages are needed. Therefore, the goal of the current work was to demonstrate the application of SPME combined with gas chromatographic separation on cyclodextrin stationary phases for the analysis of volatile chiral compounds in alcoholic beverages.

EXPERIMENTAL

Chemicals

All chemicals were purchased commercially (Aldrich Chemical Co., Milwaukee, WI or Fluka, Milwaukee, WI) and were of the highest grade available.

Gas Chromatographic Conditions

All analyses were performed using a Hewlett Packard 5890 GC and ChemStation software. Chromatographic conditions are described in Table I.

Headspace SPME

Headspace SPME sampling was done manually using a polyacrylate fiber (Supelco Inc., Bellefonte, PA). Samples (10 mL) were placed in 15 mL glass headspace vials fitted with Teflon crimp seal closures. Sodium chloride (1 g/10 mL) was added to the samples prior to sealing the vial. Following conditioning according to manufacturers directions, the SPME fiber was inserted into the headspace of the vial and allowed to equilibrate at room temperature (27°C). The actual equilibration time was dependent on the sample (22 – 68 min) and is listed on the chromatogram for each sample. Following equilibration, the fiber was injected into the inlet of the GC.

Peak identifications were determined by comparison of GC-FID retention times to those of authentic standards. Comparison of GC-MS spectra for samples and authentic standards was used for final peak confirmation. Whenever possible, absolute configuration was determined by comparison of retention times and spectra with those of enantiomerically pure reference standards purchased commercially. GC-MS conditions were the same as those used for GC-FID analyses.

Table I. Gas chromatographic conditions for analysis of flavor volatiles.

Column	CycloSil-B, 30 m x 0.25 mm x 0.25 μm (heptakis (2,3-di-O-methyl-6-O-t-butyldimethylsilyl)-β-cyclodextrin) (J&W Scientific, Folsom, CA)
Carrier Gas	Helium, 37 cm/sec measured at 40°C, or Hydrogen, 30 cm/sec measured at 40°C
Inlet	Split (1:5), 250°C (Optic Split Injector, Ai Cambridge Ltd., Cambridge, England)
Detector	FID, 250°C or HP 5973 MSD, 280°C transfer line
Oven	Oven program conditions were dependent on the sample being analyzed and are given in sample chromatograms

RESULTS AND DISCUSSION

Determination of 2-butanol and 2,3-butanediol

The stereochemical analysis of 2-butanol and its precursor 2,3-butanediol in alcoholic distillates has been used to determine the presence of spoilage bacteria during fermentation or distillation (*31 - 34*). Solvent extraction or multidimensional gas chromatography were required for the enantiomeric analysis of these compounds.

Using SPME we directly analyzed 2-butanol in distillates and 2,3-butanediol in wine (Figures 1 and 2). A polyacrylate (PA) fiber with 30 minute equilibration times (room temperature) was used. Addition of NaCl (1.5 g) was shown to improve response for these compounds

The limit of detection (LOD) was estimated to be approximately 12 μg/mL and 50 μg/mL for each isomer of 2-butanol and 2,3-butanediol, respectively, using the SPME procedure. These levels are below those typically reported in distilled beverages and wines (*31, 32*). The SPME procedure was rapid, sensitive, and can be easily automated making it ideal for routine monitoring of these compounds as an indication of bacterial spoilage of mashes or wines used for distillation.

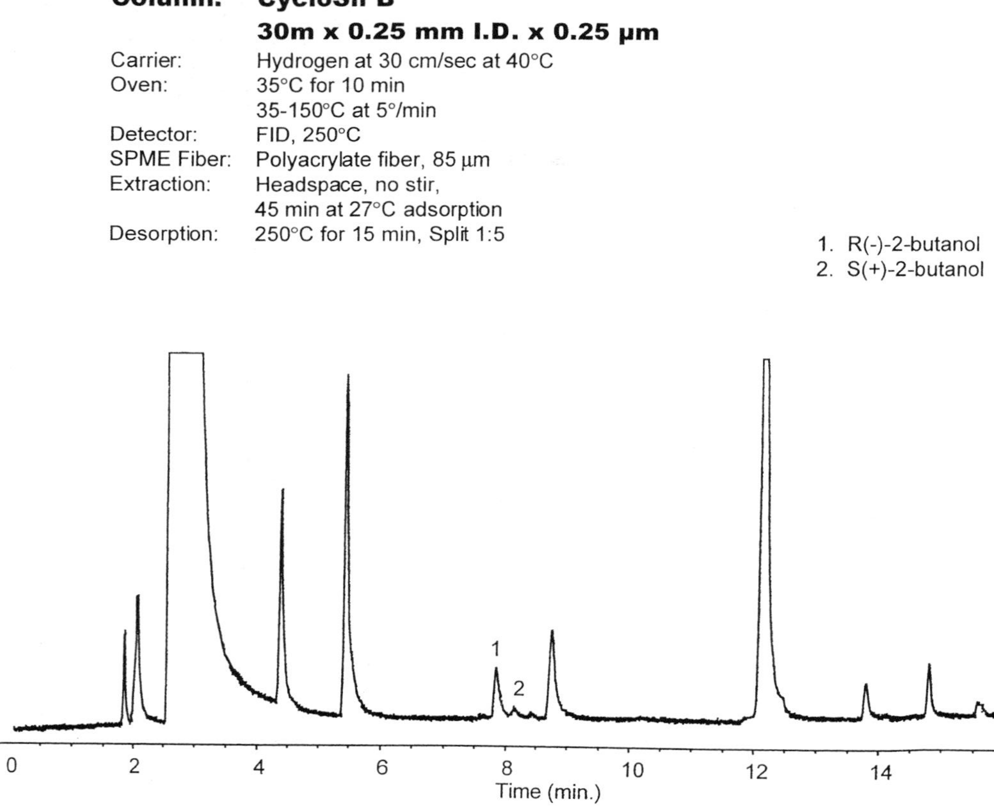

Figure 1. GC-FID chromatogram of 2-butanol in brandy.

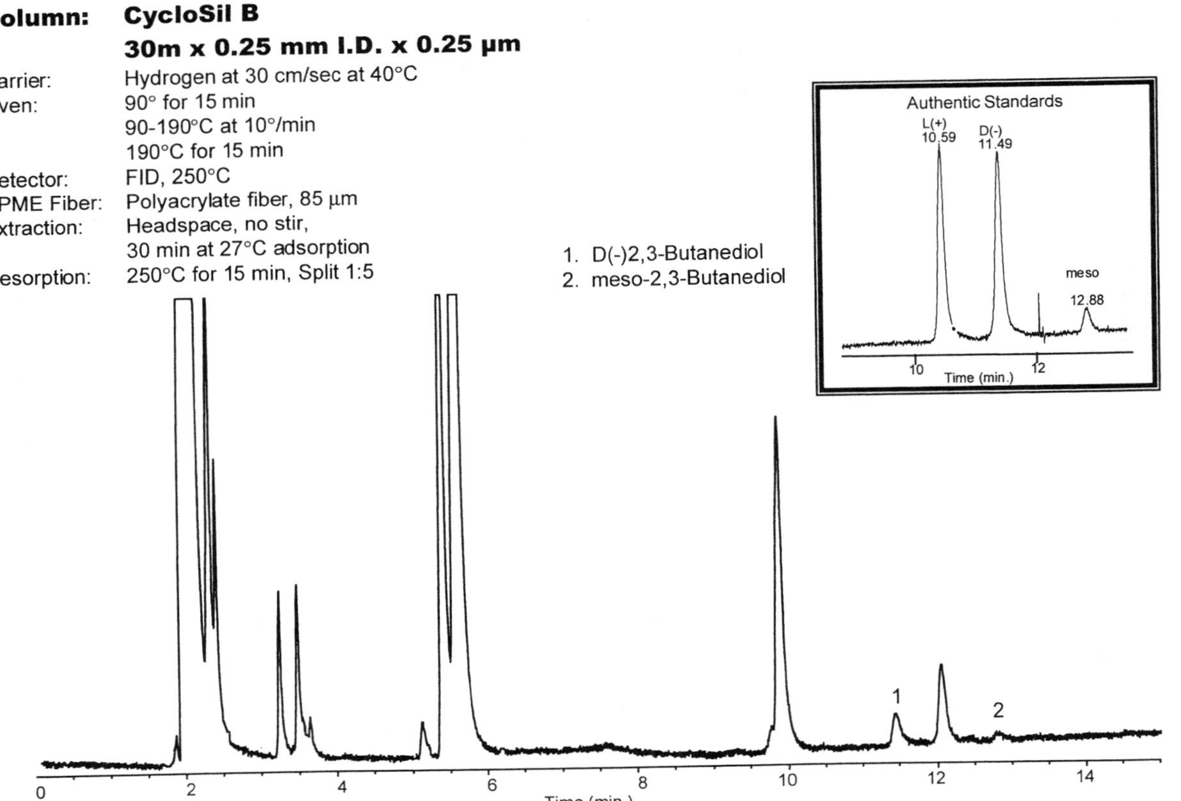

Figure 2. GC-FID chromatogram of 2,3-butanediol in white wine.

Analysis of terpenes in flavored beverages

The enantiomeric ratio of key flavor compounds has been used as a criterion for differentiating "natural" flavor compounds from their synthetic counterparts (*1, 35*). Using SPME, we analyzed flavor volatiles in a "naturally" citrus flavored malt beverage (wine cooler) and in a "natural" citrus flavored carbonated soda (Figures 3 and 4).

Several key terpenes were identified with enantiomeric compositions that were consistent between the two beverages (Table II). For example, we identified limonene which is an important constituent of citrus-based flavors, and determined that the R-(+)-isomer was predominant in both the wine-cooler and in the soda in agreement with published data for natural citrus flavors (*36*).

Care must be taken when interpreting information on enantiomeric ratios of acidic foods and beverages, however, because racemization of some compounds readily occurs (e.g., linalool, α-terpineol) (*1, 37*). Therefore, further work is needed to establish reliable criteria for distinguishing natural and synthetic flavorings in foods and beverages. However, our results show that SPME may provide a useful approach for screening foods to ascertain "quality" parameters such as natural source or origin of the flavoring ingredients.

Table II. Isomer ratios for chiral terpenes identified in a carbonated soda and a citrus-flavored malt beverage.

	Isomer Ratio[1]	
Compound	*Wine Cooler Malt Beverage*	*Carbonated Soda*
Limonene	0.10[2]	0.04[2]
Linalool	3.37	3.34
Terpinene-4-ol	0.5	0.8
α-Terpineol	1.82	1.64

[1]Isomer ratio = peak area first eluting isomer/peak area second eluting isomer

[2]Isomer ratio = S(-)-limonene/R(+)-limonene

SUMMARY AND FUTURE RESEARCH NEEDS

SPME is a promising tool for the rapid monitoring of important chiral aroma compounds. When combined with separations on cyclodextrin-based stationary phase, SPME may provide a useful approach for screening food flavorings. However, further studies are needed to improve our understanding of the incidence, chemistry, and sensory properties of chiral flavors. This information will make it possible to establish reliable criteria for evaluating "quality" parameters associated with the flavoring ingredients such as processing or storage conditions, natural source and origin.

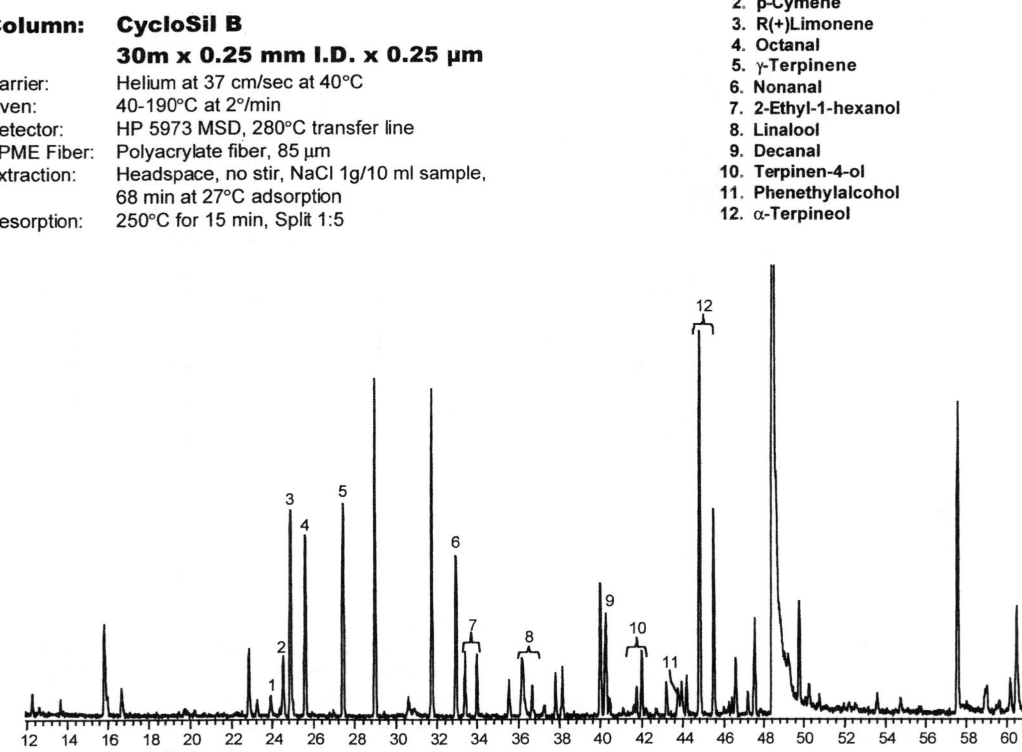

Figure 3. GC-MSD chromatogram of a citrus flavored malt beverage (wine cooler).

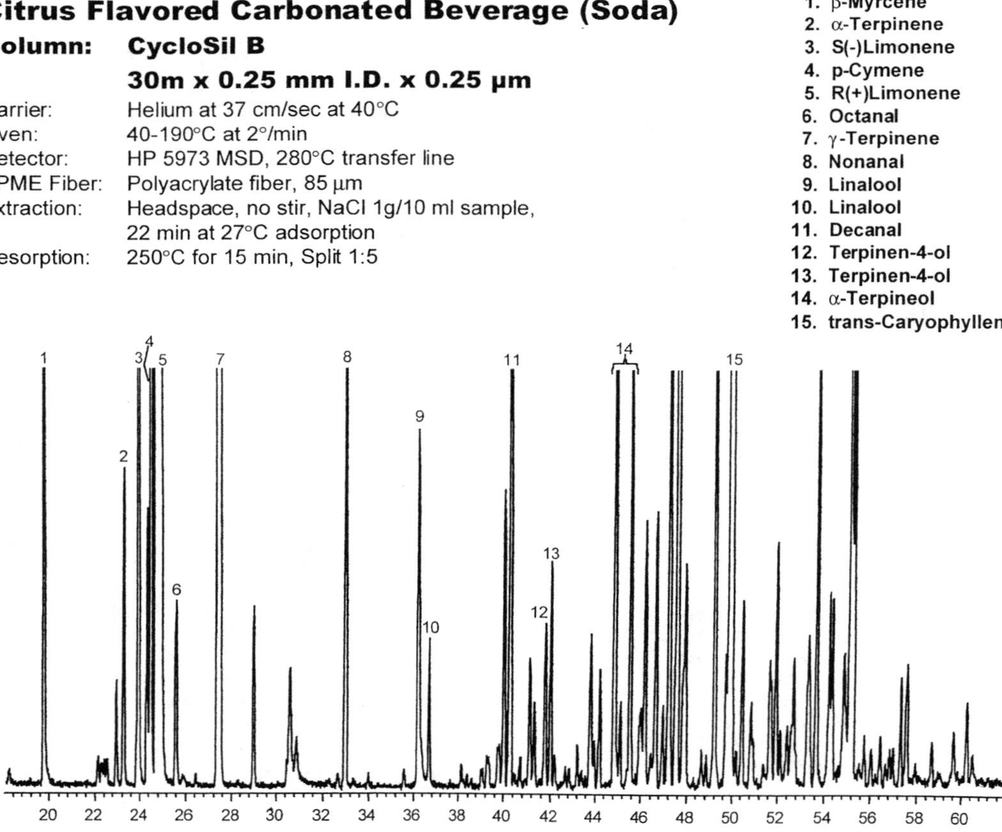

Figure 4. GC-MSD chromatogram of a citrus flavored carbonated beverage.

However, an improved understanding of the mechanisms associated with chiral separations on cyclodextrin stationary phases is also needed. Several authors have noted that small changes in analyte structure and functionality can drastically alter enantiomeric separations (*7, 9, 16, 24*). It is therefore difficult to predict the enantioselectivity of derivatized cyclodextrin stationary phases for a given analyte, and column choice is largely governed by a trial and error process. Powerful computer-based molecular modeling tools are now available which provide the opportunity to evaluate the numerous factors which can influence formation of complexes between cylcodextrins and chiral analytes (e.g., molecular size, shape, polarity, etc.). This information, when combined with experimental studies of the relationships between analyte structure and resolution on cyclodextrin stationary phases, will make it possible to better understand the chemical and physical parameters which can influence chiral recognition and chromatographic separation.

ACKNOWLEDGEMENTS

Special thanks to Jeanee Tollefson for technical assistance.

LITERATURE CITED

1. Marchelli, R.; Dossena, A.; Palla, G. *Tr. Food Sci. & Technol.,* **1996**, *7*, 113-119.
2. Casanova, Jr., J.; Corey, E. J. *Chem. Ind. (London),* **1961**, 1664.
3. Bernreuther, A.; Epperlein, U.; Koppenhoefer, B. In: *Techniques for Analyzing Food Aroma*; Marsili, R., Ed.; Marcel Dekker, Inc.: New York, 1997, pp. 143-208.
4. Gil-Av, E.; Feibush, B.; Charles-Sigler, R. In: *Proceedings 6th International Symposium on Gas Chromatography and Associated Techniques, Rome 1966;* Littlewood, A. B., Ed.; Inst. of Petroleum: London, 1967, p. 227.
5. König, W. A. *J. High Res. Chromatogr.,* **1982**, *5*, 588-595.
6. König, W. A.; Krebbber, R.; Wenz, G. *J. High Res. Chromatogr.,* **1989**, *12*, 641-644.
7. Nowotny, H.-P.; Schmaizing, D.; Wistuba, D.; Schurig, V. *J. High Res. Chromatogr.,* **1989**, *12*, 383-393.
8. Mosandl, A.; Hener, U; Hagenauer-Hener, H; Kustermann, A. *J. High Res. Chromatogr.*, **1989**, *12*, 532-536.
9. Takeoka, G.; Flath, R. A.; Mon, T. R.; Buttery, R. G.; Teranishi, R.; Güntert, M.; Lautamo, R.; Szejtli, J. *J. High Res. Chromatogr.,* **1990**, *13*, 202-206.
10. Weber, B.; Maas, B.; Mosandl, A. *J. Agric. Food Chem.*, **1995**, *43*, 2438-2441.
11. Pierce, K.; Mottram, D. S.; Baigrie, B. D. In: *Chemical Markers for Processed and Stored Foods*; American Chemical Society: Washington, DC, 1996, pp. 70-76.
12. Schmidt, G.; Neugebauer, W.; Winterhalter, P.; Schreier, P. *J. Agric. Food Chem.*, **1995**, *43*, 1898-1902.
13. Haering, D.; Koenig, T.; Withopf, B.; Herderich, M.; Schreier, P. *J. High Res. Chromatogr.*, **1997**, *20*, 351-354.

14. Dollman, B.; Full, G.; Schreier, P.; Winterhalter, P.; Güntert, M.; Sommer, H. *Phytochem. Anal.,* **1995**, *6*, 106-111.
15. Guth, H. *Helv. Chim. Acta*, **1996**, *79*, 1559-1571.
16. Jung, M.; Mayer, S.; Schurig, V. *LC-GC*, **1994**, *12*, 458, 460, 462, 464, 466.
17. Teranishi, R.; Flath, R. A.; Sugisawa, H. *Flavor Research—Recent Advances;* Marcel Dekker, Inc.: NY, 1981, 381 pp.
18. Maarse, H. *Volatile Compounds in Foods and Beverages;* Marcel Dekker, Inc.: NY, 1991, 764 pp.
19. Maarse, H.; Van der Heij, D. G. *Trends in Flavour Research: Proceedings of the 7th Weurman Flavour Research Symposium, Noordwijkerhout, The Netherlands, 15-18 June, 1993;* Elsevier: NY, 1993, 516 pp.
20. Acree, T. E.; Teranishi, R. *Flavor Science: Sensible Principles and Techniques;* American Chemical Society: Washington, DC, 1993, 351 pp.
21. Yang, X.; Peppard, T. *J. Agric. Food Chem.,* **1994**, *42*, 1925-1930.
22. Yang, X.; Peppard, T. *LC-GC*, **1995**, *13*, 882-886.
23. De la Calle Garcia, D.; Magnaghi, S.; Reichenbacher, M.; Danzer, K. *J. High Res. Chromatogr.*, **1996**, *19*, 257-262.
24. De la Calle Garcia, D.; Reichenbacher, M.; Danzer, K. *J. High Res. Chromatogr.*, **1998**, *21*, 373-377.
25. Ebeler, S. E.; Sun, G.; Datta, M.; Lautamo, R. In: *Proceedings of the 21st International Symposium on Capillary Chromatography and Electrophoresis, June 20 – 24, 1999, Park City, Utah*; 1999, p. 153.
26. Arthur, C. L; Pawliszyn, J. *Anal. Chem*., **1990**, *62*, 2145-2148.
27. Evans, T. J.; Butzke, C. E.; Ebeler, S. E. *J. Chromatogr., A*., **1997**, *786*, 298-298.
28. Fisher, C.; Fischer, U. *J. Agric. Food Chem*., **1997**, *45*, 1995-1997.
29. Hayasaka, Y.; Bartowsky, E. J. *J. Agric. Food Chem.,* **1999**, *47*, 612-617.
30. Scarlatta, C. S.; Ebeler, S. E. *J. Agric. Food Chem.*, **1999**, *47*, 2505-2508.
31. Hagenauer-Hener, U.; Henn, D.; Dettmar, F.; Mosandl, A.; Schmitt, A. *Deutsch. Lebensm.-Rundsch*., **1990**, *86*, 273-276.
32. Manitto, P.; Chialva, F.; Speranza, G.; Rinaldo, C. *J. Agric. Food Chem.,* **1994**, *42*, 886-889.
33. Speranza, G.; Corti, S.; Fontana, G.; Manitto, P.; Galli, A.; Scarpellini, M.; Chialva, F. *J. Agric. Food Chem.,* **1997**, *45*, 3476-3480.
34. Herold, B; Pfeiffer, P.; Radler, F. *Am. J. Enol. Vitic*., **1995**, *46*, 134-137.
34. Mussinan, C. J.; Hoffman, P. G. *Food Technol.,* **1999**, *53*, 54-58.
35. Bauer, K., Garbe, D., and Surburg, H. *Common Fragrance and Flavor Materials*; VCH: Weinheim, Germany, 1990, 218 pp.
36. Rapp, A. In: *Wine Analysis, Modern Methods of Plant Analysis*; Linskens, H. F.; Jackson, J. F., Eds.; Springer-Verlag: Berlin, 1988, pp. 29-66.

Chapter 6

Sensing Systems for Flavor Analysis and Evaluation

Xiaogen Yang, Jeanne M. Davidsen, Robert N. Antenucci, and Robert G. Eilerman

Givaudan Flavors Corporation, 1199 Edison Drive, Cincinnati, OH 45216

The recent development of sensing systems, such as metal-oxide or conducting polymer based electronic noses and mass spectrometer based chemical sensors, provide valuable tools for flavor chemists for research, development, application, and quality control. These sensing systems are composed of detection devices for multivariate measurements and a chemometric computing package. The concept of sensing systems can be further extended to many analytical techniques involving multivariate measurements. In the present study, some characteristics of a metal oxide based electronic nose system were evaluated, and the possibilities using other sensing systems based on UV, MS and GC/FID for discriminant analysis were explored. Applications and the potential capabilities of various sensing systems for flavor analysis and evaluation were discussed.

Introduction

Sensors are detection devices which respond to chemical or biological species in the sample and provide continuous signal output. Higher-order sensors have more than one transduction principle in the same selective layer, while chemical sensing arrays have many selective layers using the same transduction principle (*1*). Sensors can be loosely categorized into biosensors and chemical sensors based on their selectivity. Biosensors have biologically derived selectivity and chemical sensors are made of synthesized selective matrices. In the applications of sensors, especially higher-order sensors and sensing arrays, multivariate measurements are often required. The complexity of information content makes the data processing and interpretation very difficult or even impossible without the aid of statistical techniques. Therefore, multivariate statistical analysis often becomes an indispensable part of analytical processes involving sensing techniques. Chemical sensing systems are analytical systems incorporating sensors (detectors) or sensing arrays, sample introduction and/or separation (e.g., headspace sampling, chromatography), and advanced data processing techniques (*2*).

In many specific applications, a chemical sensor should ideally be highly selective, responding only to a single analyte or a certain group of chemical species and maintaining minimal interference from other species in the sample. Today, the development of new sensors with better selectivity still remains one of the major tasks in the chemical sensing research field. For flavor analysis and evaluation, however, active odor components often involve a wide range of chemical species. Therefore, it is practically impossible to develop individual sensors responding to individual odor active compounds for each application. A new approach, "Chemical Image" analysis, has been adapted recently for flavor analysis and evaluation (*3*). In this approach, the focus is on the chemical patterns characterizing an individual or group analytical subjects, rather than on the quantification and identification of each individual chemical component. Therefore, highly selective sensors for specific components are not required for this technique.

Odor recognition by the mammalian nose is achieved by a set of odor receptors with broad and overlapping selectivity. The chemical compounds that compose an aroma interact with these receptors eliciting signal response patterns that are transmitted to the brain. The brain processes the signals, conducts pattern recognition analysis, and forms a sensory image of the sample aroma (the sensory profile). This analytical principle can be applied to instrumental analysis. In general, any given analytical sample has a characteristic "chemical image" ("chemical pattern" or "fingerprint") which can be recorded using instrumental techniques. Such "chemical images" can be any collection of physicochemical, spectroscopic, and/or compositional properties of the sample. Classification and discrimination of the samples can be achieved by multivariate analysis of the "chemical image." We can also consider this method as determining the position of the sample in a multidimensional space defined by the analytical measurements. The measurements of "overall impression" may, in many cases, greatly simplify the analytical procedure. Therefore, it allows for developing a simpler, more rapid analytical method, if separation, which is generally the most time consuming process, is no longer needed.

One of the main tasks of flavor analysis is to provide qualitative and quantitative information on the flavor composition of a sample. The major analytical activities in the flavor industry are identification and quantitation of aroma components in samples. However, compositional information is not always necessary for evaluating flavor samples. For example, it is much easier to differentiate apples from oranges according to their odor, taste, color, or appearance, rather than by the analysis of the chemical composition of the fruits. In many cases, the primary analytical objective is not to provide the sample composition, but to answer questions such as: "Does this sample belong to category A or B?" or "Does this sample pass the given specification?" We can achieve these analytical objectives by recording the "chemical image" or "overall impression" of the samples using instrumental techniques, followed by pattern recognition analysis. Electronic noses are examples of such analytical instruments based on the concept of "Chemical Image."

Electronic noses are composed of an array of gas sensors to generate patterns for classification and discrimination of sample groups. Consequently, headspace sampling is required. Most of the gas sensors used in commercially available electronic noses are based on metal oxides, conducting polymers, acoustic and optical devices, quartz resonator, etc. Data analysis is often performed using statistical techniques. The gas sensor array based sensing systems have been applied successfully for discriminating complex mixtures such as foods and beverages (*4*). In some cases, a correlation between the response of the human nose and the pattern obtained from electronic noses can be established, and the information can provide objective means to evaluate odor qualities (*5*). However, this happens merely by chance and that useful correlation is just as likely not to occur.

The concept of using a sensor array to obtain a chemical image of an analytical sample can be easily extended to other types of detectors. For example, the analysis of a sample using mass spectrometry provides a characteristic chemical profile of the mixture. This mass spectrum can be utilized for classifying or for discriminating sample groups with the help of pattern recognition analysis (*6, 7*). A new sensing system based on mass spectrometry, the so-called "Chemical Sensor," has been introduced recently. This instrument is composed of a quadrupole mass selective detector (MSD) and a headspace sampling device as used in headspace GC/MS analysis. The mass spectrum of a sample headspace is recorded and the obtained ion peaks are subject to pattern recognition analysis. Because the operation and the results presented by the "Chemical Sensor" are similar to the sensor array-based electronic noses, the chemical sensor sometimes is also referred as "Mass spectrometry-based electronic nose."

The concept of "Chemical Image Analysis" can be further extended to many existing analytical techniques. For example, the MSD based chemical sensor can be viewed as GC/MS system without a GC column. In the same way, an HPLC/PDA system can be converted into a "PDA sensing system" by replacing the HPLC column by empty tubing. The obtained "chemical image" from the PDA sensing system is a UV spectrum of the sample. A method based on non-composition "chemical images" can be treated as a black box, which simplifies the analytical procedure and data analysis to a great extent. We still need to dig into the root of the cause, the chemical composition, in order to answer the question "why is the odor of apple different from orange?"

In the present study, we evaluated some characteristics of a metal oxide-based electronic nose system, and explored the possibilities using sensing systems based on UV, MS and GC/FID for pattern recognition analysis.

Experimental

Sample Preparation. All chemicals used in this study were obtained from commercial sources. The same sample preparation procedure was used for both E-nose and Fast GC analyses. 100 μL of each compound were accurately weighed into a 10 mL headspace vial and immediately capped. For very volatile flavor components,

100 μL of each chemical were dispensed into a pre-capped, tared 10-mL headspace vial, then accurately weighed. Following capping of the vial, all samples were allowed to equilibrate overnight at room temperature.

Electronic Nose System. All "electronic nose" measurements were carried out using a Fox 4000 system (Alpha M.O.S.). The Fox 4000 was equipped with a CTC HS50 Headspace Autosampler, and the Alpha M.O.S. Model 701 Air conditioning Unit. The electronic nose instrument was equipped with three sensor chambers, each containing six metal oxide sensors of various types. A flow rate of 300 mL/min of humidified air was maintained through the sensor chambers. Data was collected for 120s following the injection of a sample with a data sampling rate of 1 Hz. A 13-minute delay between the end of data acquisition and the next sample injection was allowed for the instrument equilibration.

Fast GC and MS sensing system. A Hewlett Packard 6890 GC was equipped with an FID detector and a CTC Combi-Pal Headspace Autosampler. A 10-m, 0.1-mm i.d., 0.1-μm film thickness DB-1 capillary column (J & W Scientific) was employed with hydrogen as carrier gas at a linear velocity of 40 cm/s. The oven temperature was programmed as follows: 40 °C held for 0.1 min, increased to 200 °C at a rate of 50 °C/min. The final temperature was held for 0.2 minutes, resulting in a 3.50 minute total run time.

For external calibration, standard solutions were made for each flavor chemical, using acetone or ethanol as the solvent.

The same GC was used as an MS sensing system. The injection port was directly connected to the MSD via a one-meter length of uncoated silica tubing. The injection port temperature was 250 °C and the GC oven was kept at the same temperature. The split ratio was 1:100 and the injection volume was 1 μL. The mass spectra (m/z 50 to 300) were averaged over the sample peak.

Headspace Sampling Conditions. The headspace sampling conditions were identical for both the E-nose and GC analyses. All samples were thermostatted at 35 °C for 30 minutes, with agitation. The autosampler syringe temperature was held at 40 °C.

PDA sensing system. An HP 1100 HPLC instrument equipped with PDA detector was used for the experiments. The HPLC column was replaced with empty tubing (0.12 mm i.d.). Methanol and water (75:25) were used as solvents. The flowrate was set at 0.5 mL/min. 20 μL Beverage sample was injected into the system. UV spectra were acquired from 190 nm to 400 nm. For statistical analysis, data points of the UV spectra at peak maximum were taken every 4 nm.

Statistical Analysis. The electronic nose data were processed using the built-in software package of the instrument. All other statistical analyses were performed using SPSS statistical software.

Results and Discussion

Characteristics of Electronic Noses

To study the sensitivity and selectivity of an electronic nose based on metal oxide semiconductors to common flavor components, we analyzed four homologous straight carbon chain series: alcohols, aldehydes, esters and ketones. Sensor responses were recorded at four different levels for each analyte. Comparison of the sensor responses of different analytes at identical headspace injection volumes reveals differences in both magnitude and pattern. However, the compounds analyzed also exhibit a wide range of volatility. To eliminate the volatility effect, we determined the sample headspace concentration by GC/FID under the same headspace conditions, and the injected sample amount was calculated by external calibration. The relationships between the response of the electronic nose and the injected sample amount are shown in Figure 1.

Based on the sensor selectivity, we can divide the 18 sensors into three types: high, medium, and low selectivity. The highly selective sensors can clearly differentiate the four compound classes. They have the highest sensitivity to aldehydes and ketones, medium sensitivity to alcohols, and the lowest sensitivity to esters. The second type of sensors has limited selectivity to the four compound classes. The difference of their responses to compounds with different function groups is small, but can be statistically significant. The third type of sensors exhibits virtually the same sensitivity to all compound classes tested. Figure 1 also demonstrates that the sensors have a non-linear response as a function of sample concentration. The metal oxide sensors studied were not sensitive to C-10 molecules and larger due to their low volatility at 35 °C. The selectivity of sensors to adjacent molecules in a homologous series is very limited, although they may be distinguishable in some cases (*8*). In comparison with GC/FID analysis using a narrow-bore column, the sensitivity of the sensors tested is lower by at least 2-3 orders of magnitude. Electronic noses are a headspace based technique and therefore are not sensitive for less volatile compounds at lower thermostatting temperature. Flavor analysis of food products often requires a sampling temperature at 40 °C or below in order to avoid artifact formation. This insufficient sensitivity limits the applications of electronic noses.

MS Sensing System

An MS based sensing system called "Chemical Sensor" is available commercially. It is composed of a headspace sampling device coupled to a mass selective detector and data processing software. The sample headspace vapor is directly introduced to a mass selective detector without chromatographic separation. The obtained mass spectrum of the mixture is subjected to statistical analysis.

A similar system, comparable to the commercial instrument, can be established using a typical GC/MS system by connecting the GC injection port to the MS detector with a short piece of deactivated silica tubing instead of a coated GC column. This

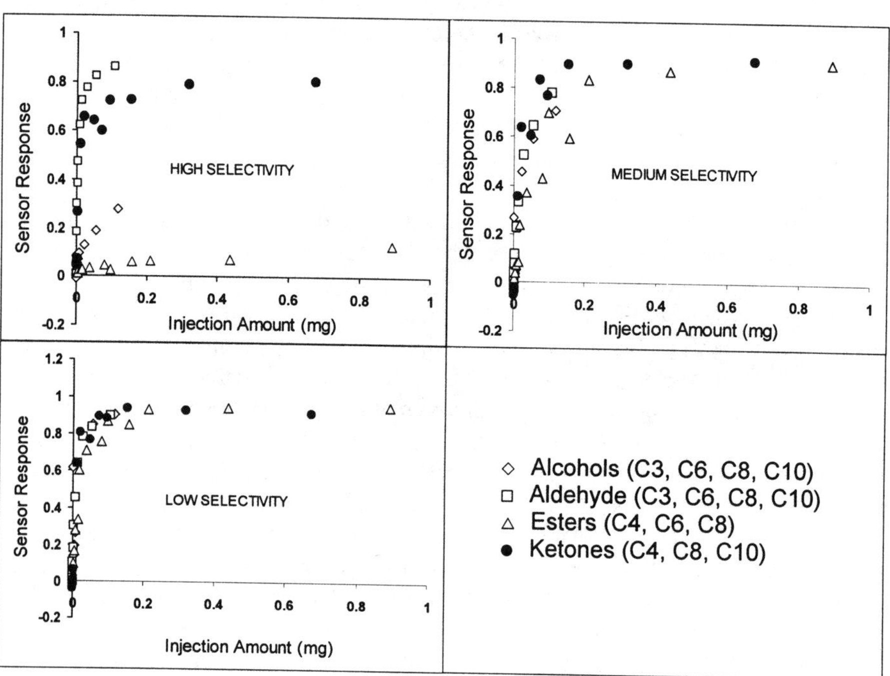

Figure 1. The sensitivity and selectivity of the Electronic Nose.

provides an additional advantage over the commercial Chemical Sensor: in addition to headspace introduction, liquid samples suitable for GC analysis can also be analyzed by the MS sensing system. The instrumental analysis can be very fast. For example, one sample per minute can be easily achieved for liquid injections, because the analytical speed is limited only by the injection speed of the autosampler. A user-developed macro program was used to extract the MS data from the stored data files and arrange it into a format suitable for statistical analysis.

In order to differentiate the aroma of salsa samples, an electronic nose and an MS sensing system were used for analysis. The analytical objective was achieved using both instruments independently. Figures 2 and 3 show the results of the analysis of salsa samples using both the metal-oxide sensor based Fox 4000 (Figure 2) and an MS sensing System (Figure 3).

Photo Diode Array (PDA) sensing system

An HPLC/PDA system performs chromatographic separation and records UV spectra of sample components. If we replace the HPLC column using a length of empty tubing, the HPLC/PDA becomes a flow injection – UV analysis system, or a "PDA sensing system." The UV spectrum of the sample is then used for pattern recognition analysis. This system is particularly suitable for liquid samples. Because the analytical time depends only on the autosampler injection time, a large number of samples can be rapidly analyzed. Beverage samples, fortified with three different types of lime oil, were analyzed using the PDA sensing system. Although sensory tests showed that the three beverages were not significantly different, the PDA sensing system easily differentiated the three samples (Figure 4).

Extending the concept of sensing systems

In general, any multivariate measurements can be used for classification or discrimination purposes. Since the chemical image of an analytical sample can be viewed from different aspects, a collection of physicochemical properties of a sample can provide valuable information for classification or discrimination purposes. A meaningful chemical image of any sample can be obtained using many analytical instruments. Therefore, sensing systems for classification or discrimination analysis can be easily established using the common analytical instruments, for example:

a) Systems obtaining 2-dimensional data
   - NIR, IR, UV, MS, NMR, ...
   - GC/FID, GC/ECD, HPLC/UV (fixed λ), ...
b) Systems acquiring three-dimentional data
   - GC/MS, GC/IR, LC/MS, LC/PDA, LC/NMR, ...

The sensing systems can also be divided into compositional and non-compositional based analysis. In most cases, instrument analyses without

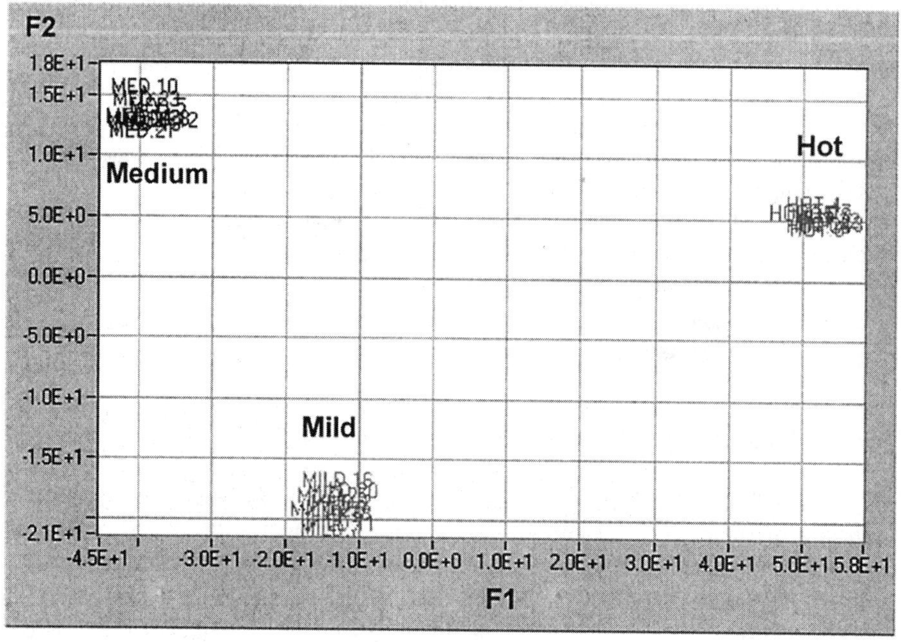

Figure 2. Discriminant Analysis of salsa samples using an electronic nose.

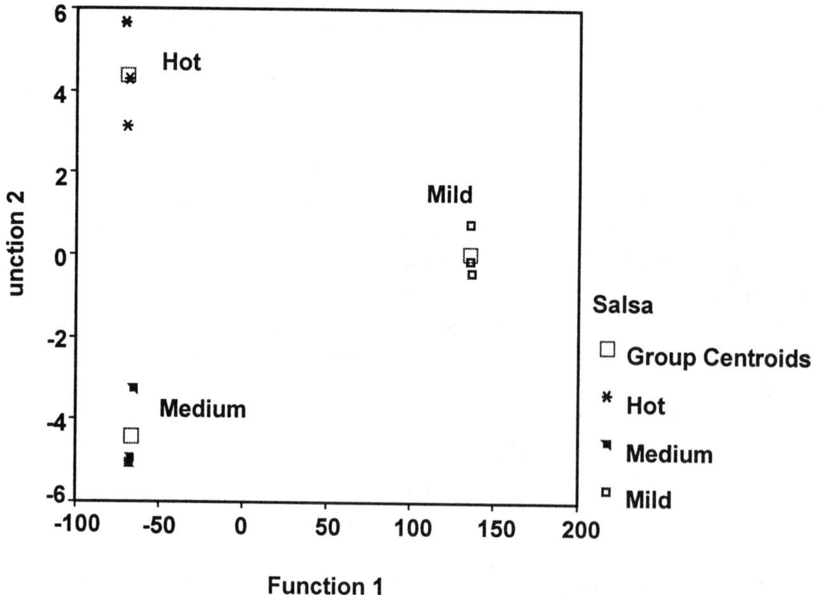

Figure 3. Discriminant Analysis of salsa samples using a MS sensor.

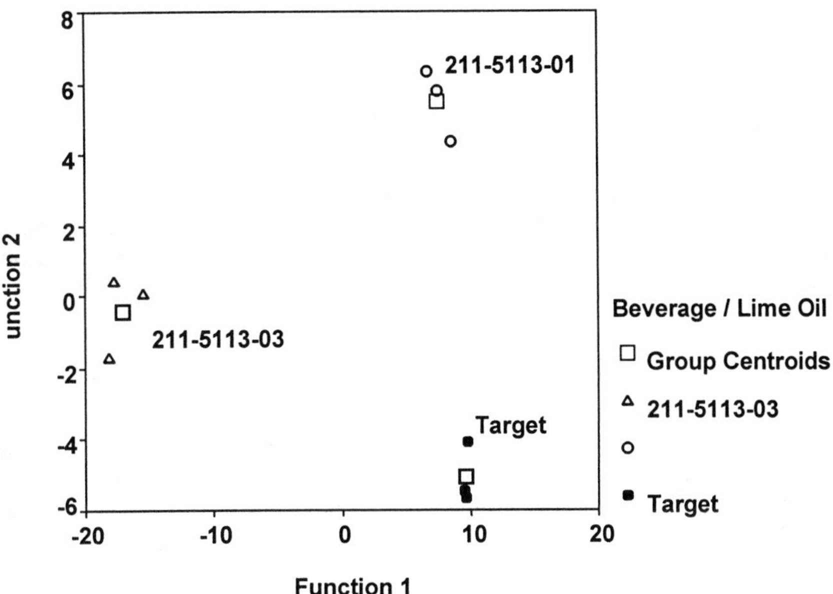

Figure 4. Discriminant analysis of beverage samples using a PDA

chromatographic separation are non-compositional based methods employing instruments or detectors such as Electronic Noses, NIR, IR, UV, EIMS, etc. These methods simplify the analytical procedure to a great extent, because the time consuming chromatographic separation is no longer needed. Therefore, they can be simple, fast, low-cost, nondestructive, and noncontacting measurements, and are very suitable for screening a large number of samples or for monitoring a production process. Although it is possible to obtain some "hints" regarding some specific components in special cases, information obtained from the non-compositional based sensing systems is limited. These measurements are normally treated as "black boxes" and it is very difficult, if not impossible, to correlate sample differences to their chemical composition without additional analysis. Sensing systems, which provide compositional information, are more valuable for the purpose of a thorough inve0stigation, and to discover the root cause of differences among samples.

GC as a sensing system

Gas chromatography (GC), a separation and detection method for multicomponent analysis of volatile compounds, is the most useful tool for flavor and fragrance analysis. GC analysis can provide qualitative and quantitative compositional information about samples. This technique has been used successfully by many investigators for classification and discrimination analysis (e.g., *9*). However, GC has not been widely used as a sensing system, especially in cases where rapid analysis of a large number of samples is required. This is mainly because GC analysis is a relatively slow process. In addition, the processing of GC data for subsequent statistical analysis can be very time consuming. Therefore, the application of conventional GC techniques is less favorable than the application of electronic noses for classification and discrimination of sample groups for reasons of analytical speed and data handling.

Although the concept of fast GC was pioneered a long time ago by chromatographic scientists, the instrumental development and application of fast GC have become popular in recent years. Using a conventional GC method, the instrumental analysis of a typical flavor sample requires about an hour. The analytical time using a fast GC can be often reduced by a factor of 10 to 20 while maintaining the chromatographic resolution. A typical GC analysis cycle (run time and oven cooling time) of flavor samples can be acomplished within 10 minutes using the fast GC technique. This analytical speed is comparable to that of a typical electronic nose analysis. In addition to the analytical speed, the GC analysis can also provide compositional information which can be utilized for further detailed analysis. With the development of personal computers and software, fast processing of GC data for statistical analysis can be achieved.

Figures 5, 6 and 7 show the results of the discriminant analysis of eight samples of an essential oil using an electronic nose and a fast GC, respectively. Figure 5 is a plot of the results of discriminant analysis using the electronic nose. This analysis reveals four, clearly differentiated groups.

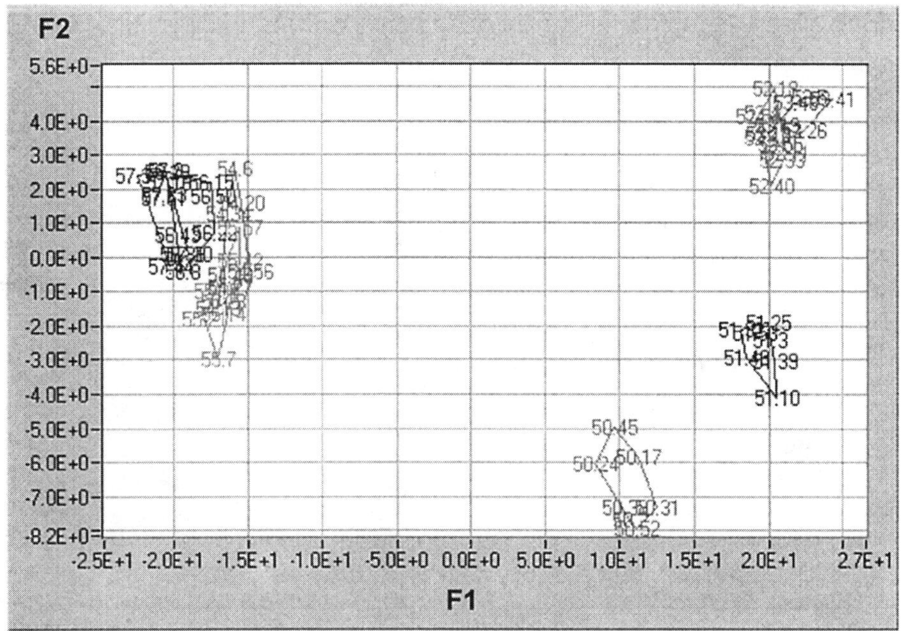

Figure 5. Discrimination of eight different samples of an essential oil by electronic nose.

The same sample group was analyzed, in triplicate, using fast GC, and GC/MS was used for identification. The peaks of the fast GC and the conventional GC/MS chromatograms were correlated by retention indices. Figures 6 and 7 show the results of the discriminant analysis. The eight samples are significantly different from each other. In the plot of function 1 *vs.* 2, all sample groups are discriminated except the groups 4 and 6. Examination of the plot of function 1 *vs.* 3 shows that groups 4 and 6 are significantly different according to discriminant function 3 (Figure 7).

One of the advantages of compositional based sensing systems is the ability to correlate the differences found among sample groups to their chemical composition. The loadings of the vectors in the discriminant analysis revealed the components that accounted for most of the variation among the samples. In this example, the analysis of the discrimant functions revealed that *trans*-ocimene is the component which contributes to the major differences among the samples on discriminant function 1 (Figure 8) and ethanol is the component which contributes to the major differences

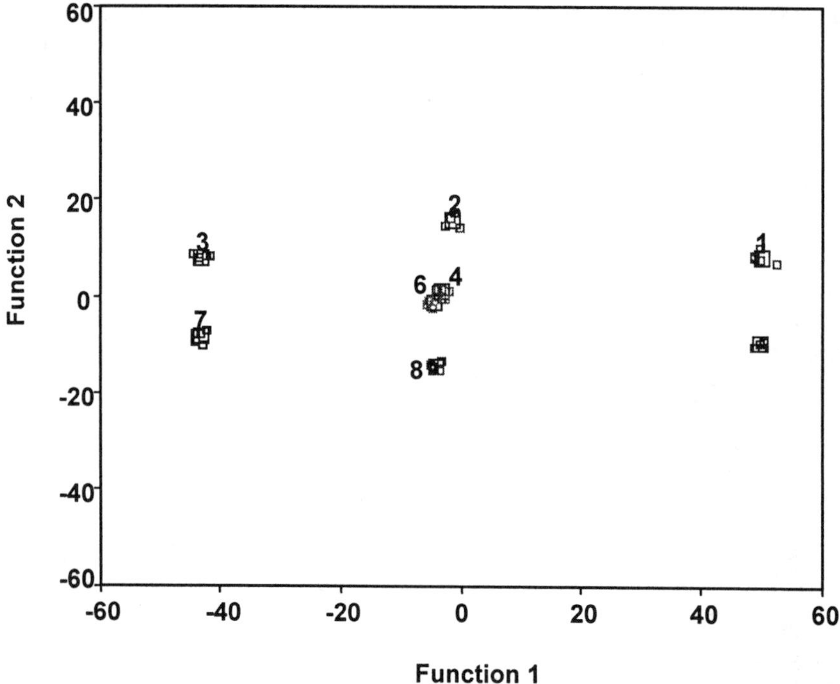

Figure 6. Discriminant analysis of eight samples of an essential oil by Fast-GC. Discriminant function 2 vs. *discriminant function 1.*

among the samples on discriminant function 2 (Figure 9). Furthermore, the investigation of concentration differences among the sample groups revealed that the peak area percentage of ethanol in samples 1, 2, 3, and 4 is significantly higher than in samples 5, 6, 7, and 8 as shown in Figure 10. This suggested that ethanol concentration was probably the major variable in differentiation of the samples. To determine the importance of ethanol as a discriminant factor, the independent variable ethanol was removed from the data set, and a second discriminant analysis was performed. The results are shown in Figure 11 as a plot of discriminant function 1 *vs.* discriminant function 2. In this analysis, no significant differences were found between samples 1 and 5, 2 and 6, 3 and 7, and 4 and 8. Ethanol is commonly used as a solvent in flavor production. The gas sensors employed in electronic noses are very sensitive to ethanol, therefore, the major difference detected by the electronic nose is probably due to different ethanol content.

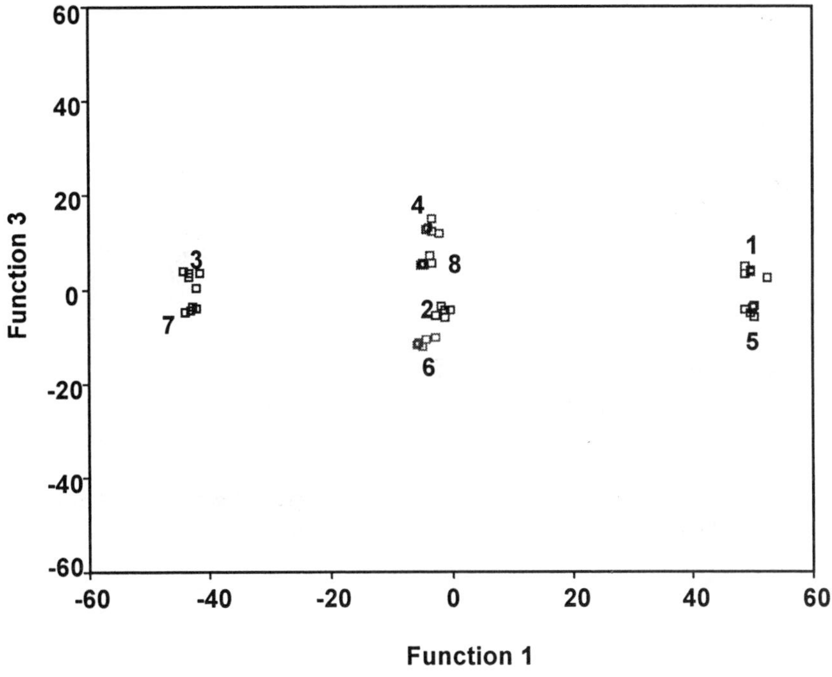

Figure 7. Discriminant analysis of eight samples of an essential oil by Fast GC. Discriminant function 3 vs. discriminant function 1.

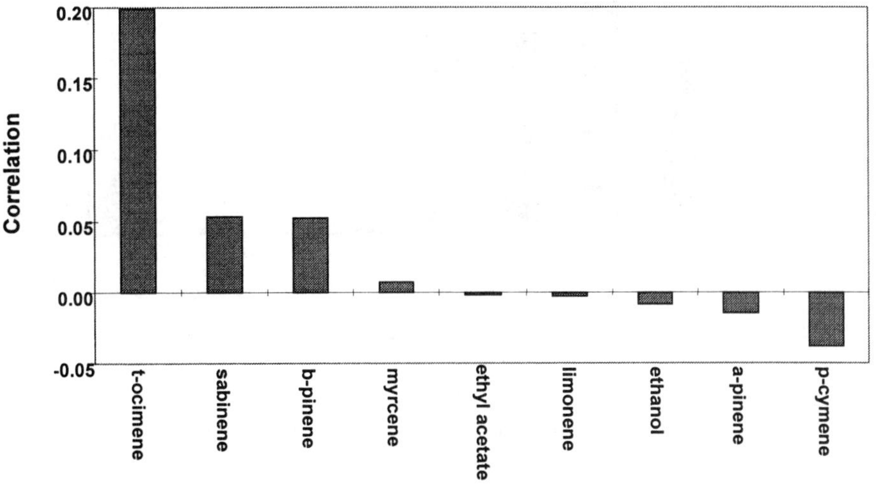

Figure 8. Discrimininant analysis of Essential Oil samples: Variable loadings for discriminant function 1.

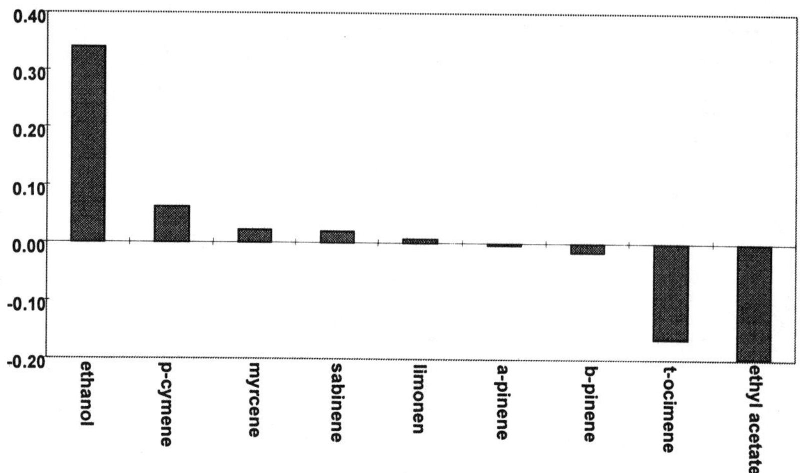

Figure 9. Discriminant analysis of Essential Oil samples: Variable loadings for discriminant function 2.

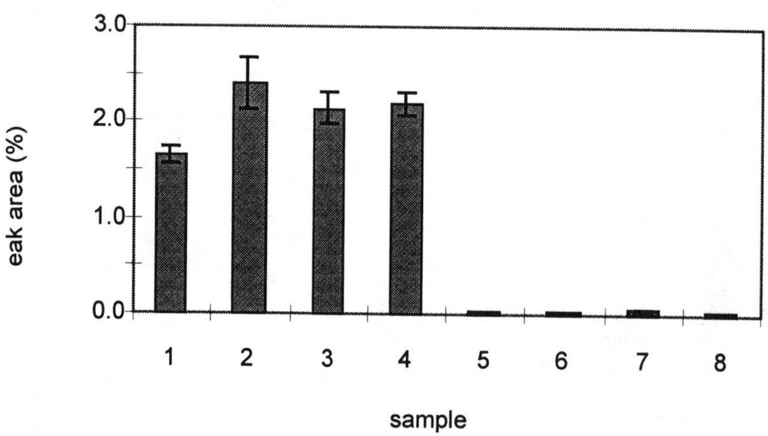

Figure 10. Ethanol percent peak area for each essential oil sample.

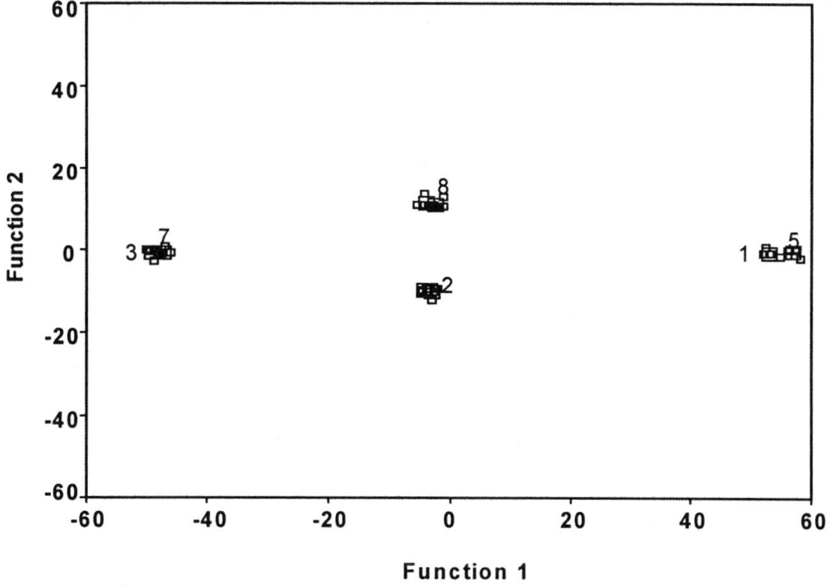

Figure 11. Discriminant analysis of essential oil samples analyzed by fast GC following the removal of ethanol from the data set.

Summary

A chemical sensing system is an analytical system for multivariate detection and analysis. These variables can include specific chemical components, or alternatively electrochemical data, spectroscopic data, or any other physicochemical properties. The most often used multivariate analysis methods include principal component analysis, discriminant analysis, cluster analysis, and artificial neural network simulation. These sensing systems generally can conduct fast analysis and provide simple and straightforward results and are best suited for quality control and process monitoring in the flavor industry. The advantage of fast GC as a sensing system is that it can perform rapid analysis but also provide quantitative compositional information. This information can be used not only for pattern recognition analysis, but also for in-depth studies.

Many attempts have been made to construct sensors and sensing arrays, which mimic the mammalian nose. A similar sensing principle to the mammalian sense for odor, as we understand so far, is relatively easy to apply to developing sensing systems. However, we still have a long way to go to develop a sensor that produces the same or similar signal pattern as the nose. The sensitivity of the mammalian nose for certain chemical species still far exceeds even the most sensitive modern analytical instruments. This should be kept in mind when correlating analytical instrumental data with sensory evaluation results.

Literature Cited

1. Janata, J.; Josowicz, M.; Vanysek, P.; DeVaney, D.M. *Anal. Chem.* **1998,** *70,* 179R-208R.
2. Wise, B. M.; Januta, J. *Encyclopedia of Energy Technology and the Environment*, J. Wiley & Sons: New York, NY, 1995; pp 1247-1258.
3. Albone, E.; In *Handbook of Biosensors and Eletronic Noses*; Kress-Rogers, E. Ed., CRC Press: Boca Raton, 1997; pp 503-519.
4. Schaller, E.; Bosset, J.O.; Escher, F. *Lebensm. Wiss. U. Technol.* **1998**, *31*, 305-316.
5. Annor-Frempong, I.E.; Nute, G.R.; Wood, J.D.; Whittington, F.W.; West, A. *Meat Science* **1998**, *50(2)*, 139-151.
6. Goodacre, R., Kell, D.B. *J. Sci. Food Agric.* **1993**, *63*, 297-307.
7. Goodacre, R., Kell, D.B., Blanchi, G. *Nature* **1992**, *359*, 594
8. Schaak, R.E.; Dahlberg, D.B.; Miller, K.B. In *Electronic Noses & Sensor Array Based Systems, Design & Applications;* Hurst, W.J. Ed.; Technomic Publishing Co.: Lancaster, PA, 1999; pp 14-26.
9. Peppard, T.L. *Food Quality and Preference* **1994**, *5*, 17-23.

Chapter 7

Novel Aspects of Tomatillo Flavor

R. J. McGorrin[1] and L. Gimelfarb[2]

[1]Department of Food Science and Technology, Oregon State University, Corvallis, OR 97331
[2]Technology Center, Kraft Foods, Inc., 801 Waukegan Road, Glenview, IL 60025

Identifications of the volatile consitutents of fresh and cooked tomatillo were performed using dynamic headspace concentration and capillary GC-MS analyses. In the present study, over 50 volatile compounds were identified in fresh tomatillo, including hydroxy esters, saturated and unsaturated aldehydes, alcohols and terpenes. A quantitative method for calculation of Odor Unit values was applied to determine the relative significance of identified volatiles and assess their sensory impact to tomatillo flavor. Data obtained from this method indicates that (Z)-3-hexenal, (*E*,*E*)-2,4-decadienal, nonanal, hexanal, hexanol and (Z)-3-hexen-1-ol were the most significant contributors to the "green" impact. In comparison with volatiles identified in plum tomato, the flavor compounds unique to tomatillo included hydroxy esters, aromatic esters, C8-C12 aldehydes, decanoic acid, and di- and sesquiterpenes.

The tomatillo or husk tomato (*Physalis ixocarpa* Brot. syn. *Physalis philadelphica* Lam.) is the major solanaceous ingredient used in the preparation of the green sauces and salsas in various Mexican dishes. The tomatillo is related to, but horticulturally distinct from, the ground cherry (*Physalis pruinosa* L.) which is cultivated in parts of Europe (*2*). It is increasingly being utilized in a variety of culinary recipes which blend and adapt Latin American flavor themes to contemporary North American tastes. While botanically classified as a fruit, the tomatillo is typically consumed in savory dishes, in the context of a vegetable. Similar only in appearance to a green tomato, the tomatillo imparts a unique flavor profile. As an additional difference, tomatillos are surrounded with a tan/green paper-like husk, which covers a sticky-waxy substance on the surface of the fruit skin.

An emerging flavor trend is the increased use of novel food ingredients to provide new culinary experiences and variety (*3*). Tomatillos are readily available as fresh vegetables in the produce section of U. S. supermarkets, since they are grown in California and other regions of the Americas. Consequently, tomatillos are increasingly being utilized in a variety of "fusion cooking" recipes. In this context, tomatillos are combined with ingredients from several Latin American dishes and heated to create unique flavor blends which are suitable for mainstream acceptance, yet which have an "authentic" appeal (*4*).

Our initial report on tomatillo flavor (*5*) compared the major flavor volatile differences between fresh tomatillo and red plum tomato. While more is known about the flavor chemistry of fresh and thermally processed tomatoes and tomato juice (*6-12*), correspondingly few published studies are available regarding the significant flavor components of tomatillo. Previous literature reports on tomatillo have limited information on only non-volatile flavor components derived from compositional data (*1*). As a continuation of our preliminary study, we are interested in isolating and identifying the volatile and non-volatile flavor components in fresh and cooked tomatillos, and to compare similarities and differences to those previously reported in tomato flavor. We have focused on applying quantitative techniques which enable identification and classification of the most potent volatile compounds with sensory significance, to derive a better understanding of the distinctive flavor of this novel food ingredient.

Materials and Methods

Materials

Samples of tomatillos were purchased from a local food supermarket in the Chicago metropolitan area of the United States. They were cultivated in the Curical, Sinaloa region of Mexico and harvested during the Spring, 1997 growing season. A sample of 4-heptanone (internal standard) was obtained from Aldrich Chemical Co. (Milwaukee, WI) and used without further purification. Tenax TA (20/35 mesh) was obtained from Chrompack (Raritan, NJ). Silanized glass wool was obtained from Supelco Inc. (Bellefonte, PA).

Sample Preparation

Fresh tomatillos (1.5 lb., approximately 15 count) were pooled, the husks were removed, the fruits were rinsed with water, and dried. After removing the stems, the samples were pureed in a Sunbeam food processor. For heat-treated samples, the purees were heated with constant stirring in an open stainless steel pan to 100 °C, held for 2 minutes, and rapidly cooled in an ice bath. Samples requiring volatile flavor measurements were analyzed by headspace gas chromatography–mass spectrometry (GC-MS) within one day of preparation.

Lipid Content and Fatty Acid Composition

Normalized fatty acid distribution was determined by methylation of the fatty acids and glyceride extracts of tomatillo purees by transesterification with sodium methoxide (*13*), followed by analysis of the resulting methyl ester derivatives using a gas chromatograph (Hewlett Packard 5890). The GC capillary column used for fatty acid analysis was DB-WAX, 30 m x 0.32 mm i.d., 0.25 μm film thickness (J&W Scientific, Folsom, CA).

Proximate and Non-Volatile Assays

Analyses of moisture, pH, and titratable acidity were performed at Silliker Laboratories (Chicago Heights, IL) using standard AOAC analytical methods. Sugars (glucose, fructose, sucrose) were assayed using an AOAC reversed phase liquid chromatography procedure (*14*). Titratable acid content was calculated as citric acid after titration with 0.1 N NaOH. Organic acids (citric, malic, lactic, oxalic, acetic, tartaric) were determined using an ion exchange chromatography method (*15*).

Volatile Component Isolation

Tomatillo purees (3.0 g) were placed in a 100-mL Envirochem (Kemblesville, PA) sparging vessel and mixed with 5 g of distilled water. A 4-heptanone internal standard (0.16 ppm based on the total weight of sample) was spiked into each matrix. Samples were purged for 45 min. at 50 °C with a nitrogen flow rate of 106 mL/min onto a glass desorption tube (3.0 mm i.d. x 16 cm length) packed with 100 mg 20/35 mesh of Tenax-TA adsorbent. Prior to thermal desorption, the tube was purged for 15 min at 50 °C with dry nitrogen gas (30 mL/min) to remove traces of moisture.

GC-MS Identifications of Volatile Components

Tenax cartridges were thermally desorbed into a Hewlett Packard GC-MS instrument via a Model 4010 Thermal Desorption System (Chrompack, Raritan, NJ). The Tenax trap was heated at 220 °C for 10 minutes and volatiles were cryofocused onto a fused silica cold-trap held at -140 °C. Volatiles were directly desorbed for 1 min at 220 °C with a helium carrier gas flow rate of 1.0 mL/min (37.5 cm/sec linear velocity) onto a cooled capillary column. The GC column was a 30 m x 0.25 mm DB-5 MS fused silica open-tubular capillary column, 0.25 µm film thickness (Hewlett Packard, Wilmington, DE). The GC oven temperature was ramped using a multi-stage temperature program starting at -10 °C for the first 3 min, increasing at 10 °C /min to 40 °C (0 min hold); then slowing to 3 °C/min to 140 °C (0 min hold); and finally 8 °C/min to 230 °C, with a 5 min hold at the upper limit. The GC column was directly interfaced to a Hewlett Packard 5972 mass selective detector via a heated transfer line maintained at 280 °C. The mass spectrometer was operated in the electron ionization mode (70 eV) scanning masses 33-350 *m/z* at 2.2 scans/sec. The electron multiplier voltage was increased 10% to provide optimal sensitivity. All mass spectra were background-subtracted and library-searched against the National Institute of Standards and Technology mass spectral reference collection. The Wiley/NBS Registry of Mass Spectra and DB-5 Kovats retention time indices (*n*-alkane standards) were used to facilitate compound identification.

Results and Discussion

Non-Volatile Composition

A summary of compositional analyses for tomatillos is presented in Table I. Data from this study includes literature values for protein, total lipid, carbohydrate and mineral assays. Sensory evaluation of fresh tomatillo provides a dominant sour taste. The perceived sourness of tomatillo is contributed by a low pH (3.83) and moderate titratable acidity (1.11% as citric acid). Acidity values increase slightly with cooking, related to moisture loss and the corresponding increase of acid concentration levels.

Quantitative data on organic acids by ion exchange HPLC analysis are presented in Table I. The predominant organic acid in tomatillo was found to be citric acid, accompanied by small amounts (< 0.06%) of malic and lactic

Table I. Composition of Selected Non-Volatile Components of Tomatillo

	Fresh	Cooked
Water *(%)*	93.0	91.2
pH	3.83	3.85
Titratable acidity *(as citric acid)*	1.11	1.23
Protein (N x 6.25) *(%)*	1.0[a]	
Lipids	0.7[a]	
Carbohydrates *(wt. %)*	4.5[a]	
- Fructose	0.9	1.6
- Glucose	1.0	0.8
- Sucrose	0.6	0.8
Organic acids *(wt. %)*		
- Citric	1.30	1.63
- Malic	0.02	0.02
- Lactic	0.06	
- Oxalic	0.03	
Carotenoids *(mg %)*	4[a]	
Minerals *(mg %)*		
Calcium	18[a]	
Iron	2.3[a]	

[a] Data obtained from reference 1.

acids. Despite the sensory impression of oxalic acid, very low quantities were measured. Acetic and tartaric acids were undetectable above the 0.01% detection limit for the HPLC method.

Sugar composition results provided some intriguing comparisons (Table I). The data show different distributions and overall higher levels of sugars in cooked vs. fresh tomatillo. The increase of simple sugars during cooking suggests the breakdown of complex polysaccharides during heating in the presence of acid. Of interest is the relatively higher levels of fructose and sucrose in cooked tomatillo. This compositional distribution of sugars should conceivably influence the generation of alternate flavor compounds for tomatillos through participation in Maillard pathways during the cooking process.

The results of the fatty acid profile comparison are presented in Table II. Of significance is that a greater proportion of the lipid fraction in tomatillo is comprised of decanoic acid (C10:0, 20.0%). The other predominant fatty acids are linoleic (C18:2, 30.1%) and palmitic (C16:0, 20.2%). The profile data for cooked tomatillo reflect a significant increase in the proportion of C18:2 in the total lipid fraction after heating, with a simultaneous decrease in C10:0. This suggests that fatty acid precursors are available for formation of flavor-significant saturated and unsaturated aldehydes in the maturation stages of the fruit and during cooking. Of interest is that tomatillos have 4-fold increases in total lipid and iron contents vs. tomato, which may have flavor precursor implications. (*16*)

Volatile GC-MS Identifications

Flavor isolation techniques such as dynamic headspace concentration and simultaneous distillation-extraction (SDE) have been commonly used for the analysis of tomato and other natural flavors (*6,10*). Volatile headspace trapping on Tenax adsorbent resins has become a mainstay for flavor analysis laboratories because it is moderately sensitive, can be performed rapidly, and is a mild sampling technique which is not prone to generate thermal artifacts, as frequently happens for SDE. Because we were interested in quantifying the effects of cooking, we selected this technique to compare flavor differences among fresh and heated tomatillos. The repeatability of the technique has been established with model mixtures, and we chose 4-heptanone as an internal standard for quantification using MS total ion peak area.

The volatiles from 3g of tomatillo samples were concentrated in approximately 5 liters of headspace through a Tenax adsorption cartridge and

Table II. Normalized Fatty Acid Distribution for Tomatillo

Fatty acid	GC area % Fresh	GC area % Cooked
10:0	20.03	11.84
12:0	2.74	1.52
14:0	0.63	0.27
14:1	1.17	0.38
16:0	20.23	18.62
16:1	1.26	0.94
18:0	2.77	2.76
18:1 ω9	8.59	9.82
18:2 ω6c,c	30.13	40.19
18:3 ω3c,c,c	8.79	8.96

desorbed directly onto a cooled DB-5 capillary column. The GC-MS total ion chromatogram (Figure 1) from the volatile fraction displays the identities of approximately 20 key compounds with flavor significance in fresh and cooked tomatillo. The GC-MS volatile profile obtained for fresh tomatillo was highly complex, as is typical for natural products, and a total of 52 volatile compounds were identified. Table III provides a more detailed list of volatile compounds identified by GC-MS, along with Kovats GC retention indices and quantitative data, which was obtained relative to a 4-heptanone internal standard. The GC retention indices enabled comparison of flavor compound profiles to those reported in the literature on DB-5 columns.

Lipid Derived Volatiles

The most predominant volatiles were saturated and unsaturated 6-carbon aldehydes and alcohols. (*Z*)-3-Hexenal and (*E*)-2-hexenal were present in significant concentrations, and contribute a green, leafy herbaceous aroma. These compounds have been previously associated with the characteristic green, tomato fruit-like aroma released by tomato leaves (*9*). The unsaturated aldehydes are formed from linolenic acid by reaction with lipoxygenase enzyme via a 13-hydroperoxide intermediate (*17*). The bioreduction products, (*Z*)-3-

Table III. Volatile Compounds Identified by Headspace GC-MS in Fresh and Cooked Tomatillo

No.[a]	RI[b] (DB-5)	Compound	Fresh (ppb)[c]	Cooked (ppb)[c]
	723	2-Methyl-2-pentenal	20	
	736	Dimethyl disulfide	tr.	
1a	753	Acetic acid		820
1	775	*iso*-Butyl acetate	240	
2	800	Hexanal	5,400	820
	815	Butyl acetate	20	
3	839	(*E*)-2-Hexenal	2,000	160
4	849	(*Z*)-3-Hexenal	21,000	3,300
5	875	(*Z*)-3-Hexen-1-ol	2,700	800
6	895	1-Hexanol	26,000	7300
7	897	Methyl 2-hydroxy-*iso*-pentanoate		320
	913	2,4-Hexadienal	110	
	932	Methyl hexanoate	25	
8	951	Benzaldehyde	540	2,200
9	952	(*E*)-2-Heptenal	"	"
	957	Dimethyl trisulfide	tr.	
	984	2-Pentylfuran	tr.	
10	986	6-Me-5-hepten-2-one	240	
11	997	Methyl 2-hydroxy-3-methyl pentanoate	820	510
	998	Octanal	110	110
	1004	(*Z*)-3-Hexen-1-yl acetate	180	
12	1011	Hexyl acetate	250	
	1014	(*E*)-2-Hexen-1-yl acetate	tr.	
13	1016	Limonene	270	300
	1033	Benzyl alcohol	tr.	
14	1054	(*E*)-2-Octenal	120	70

Table III Continued

No.[a]	RI[b] (DB-5)	Compound	Fresh (ppb)[c]	Cooked (ppb)[c]
	1079	α-Terpinolene	tr.	
	1088	Methyl benzoate	95	
	1098	Linalool	tr.	tr.
15	1102	Nonanal	2,000	710
	1114	2-Phenylethanol	tr.	
	1153	(*E*)-2-Nonenal	60	40
	1163	Ethyl benzoate	tr.	
	1167	Terpinen-4-ol	50	40
	1175	Cumic alcohol		50
	1183	Methyl salicylate	85	75
	1187	Myrtenol	tr.	
	1191	Ethyl octanoate	tr.	
16	1197	Decanal	320	300
	1231	Neral	25	
	1237	43,55,70,83,92,125	430	200
	1252	(*E*)-2-Decenal	tr.	
	1263	Geranial	60	
17	1285	(*E,Z*)-2,4-Decadienal	22	
18	1309	(*E,E*)-2,4-Decadienal	160	
	1373	Decanoic acid	tr.	
19	1374	β-Damascenone	1	3
	1400	Dodecanal	85	120
20	1443	Geranyl acetone	90	
21	1473	β-Ionone	49	
	1619	Cadinene	30	

[a] Peak numbers refer to Figure 1.

[b] Kovats retention index, relative to *n*-alkane hydrocarbon standards.

[c] Concentrations calculated from 160 ppb 4-heptanone internal standard.

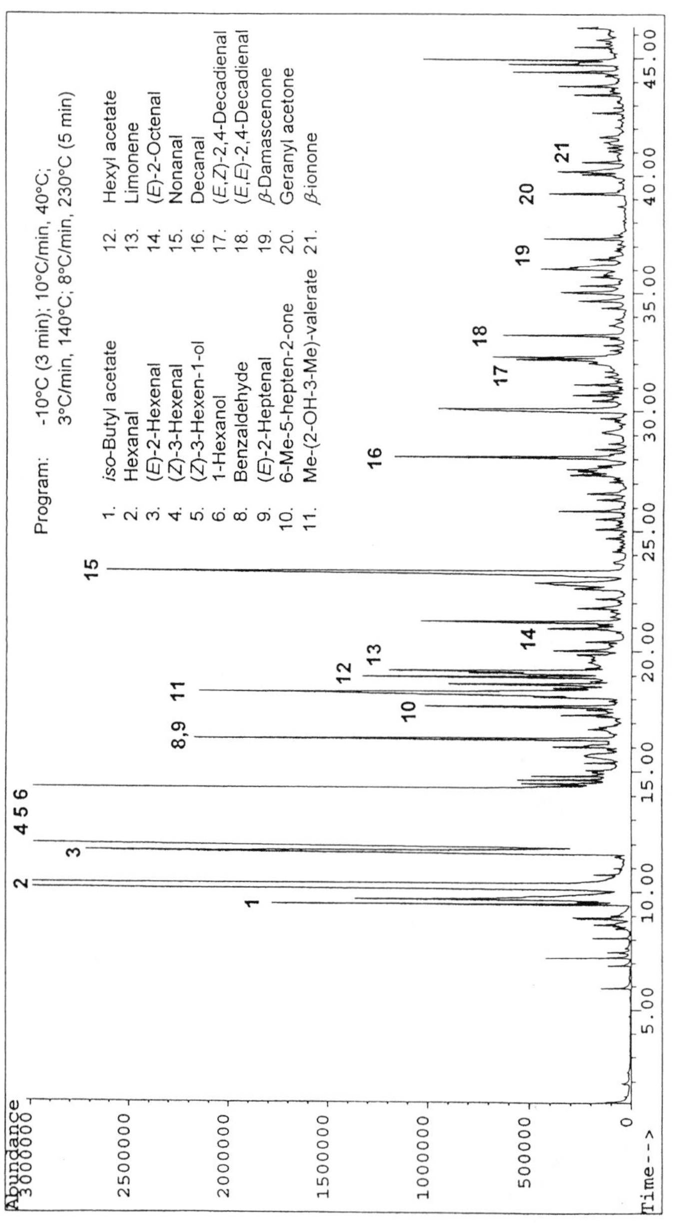
Abundance
3000000
2500000
2000000
1500000
1000000
500000
0
Time-->
5.00
10.00
15.00
20.00
25.00
30.00
35.00
40.00
45.00
Program: -10°C (3 min); 10°C/min, 40°C;
3°C/min, 140°C; 8°C/min, 230°C (5 min)
1. iso-Butyl acetate
2. Hexanal
3. (E)-2-Hexenal
4. (Z)-3-Hexenal
5. (Z)-3-Hexen-1-ol
6. 1-Hexanol
8. Benzaldehyde
9. (E)-2-Heptenal
10. 6-Me-5-hepten-2-one
11. Me-(2-OH-3-Me)-valerate
12. Hexyl acetate
13. Limonene
14. (E)-2-Octenal
15. Nonanal
16. Decanal
17. (E,Z)-2,4-Decadienal
18. (E,E)-2,4-Decadienal
19. β-Damascenone
20. Geranyl acetone
21. β-ionone
1
2
3
4 5 6
8,9
10
11
12
13
14
15
16
17
18
19
20
21

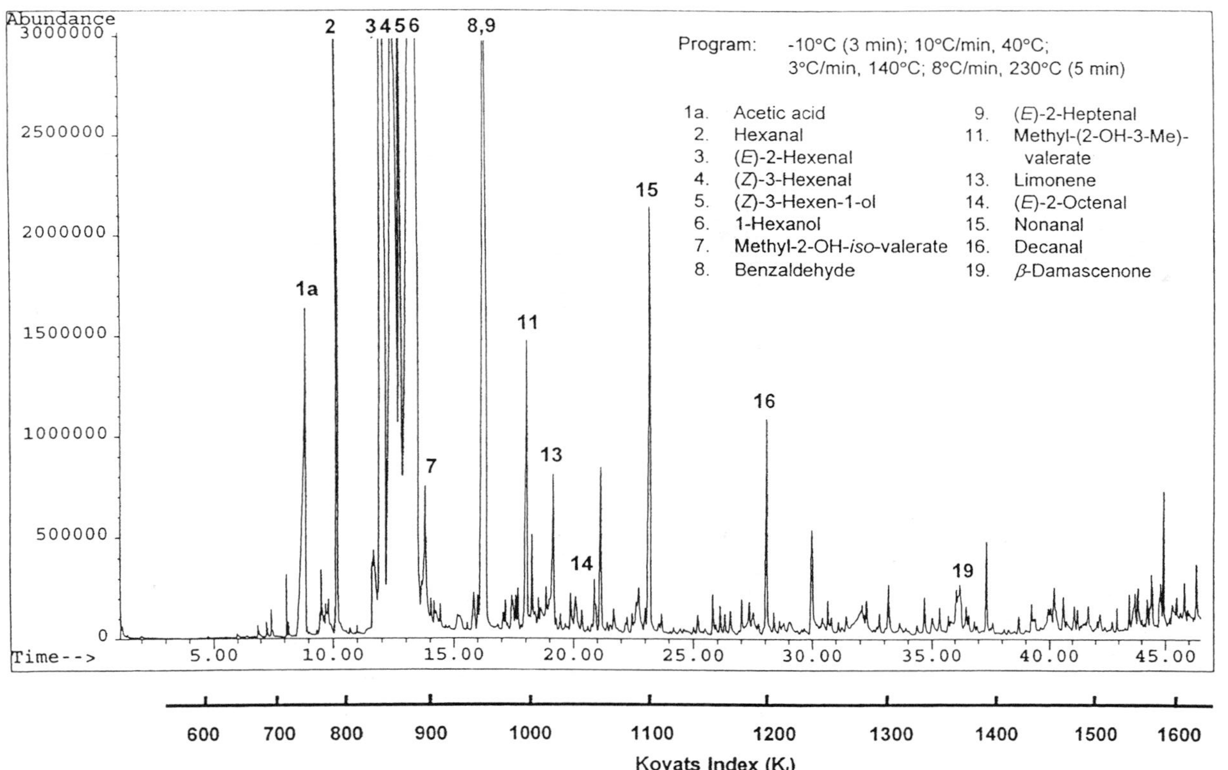

Figure 1. Comparison of fresh (top) and cooked (bottom) tomatillo flavor volatiles obtained by dynamic headspace GC-MS (DB-5).

hexen-1-ol and (*E*)-2-hexen-1-ol, were observed at lower concentrations; (*Z*)-3-hexen-1-ol was previously reported by Buttery and Ling (*9*) to form in tomato via a reductase conversion pathway.

Hexanal, one of the major aldehydes in tomatoes, is considered to be important for fresh tomato flavor (*6,8*), and is also a contributor to the green, fatty component of tomatillo aroma. Similar to the unsaturated aldehydes, it is derived from a linoleic acid breakdown pathway via lipoxygenase formation of predominantly the 13-hydroperoxylinoleic acid intermediate (*6,17*). 1-Hexanol provides a winey, cider-like character and is produced via bioreaction of hexanal with alcohol dehydrogenase. Other significant aldehydes include 8- to 10-carbon enals including 2-octenal, 2-nonenal, and 2,4-decadienal, which contribute fatty-green aromatics.

Esters

After the aldehydes, the largest volatile class of compounds identified in tomatillos were esters. Among these were butyl, *iso*-butyl, hexyl, (*Z*)-3- and (*E*)-2-hexenyl acetates, methyl hexanoate, and ethyl octanoate. Aromatic esters include methyl and ethyl benzoate, and methyl salicylate. Two unique hydroxy esters were identified by matches with the mass spectral library: methyl 2-hydroxy-3-methylpentanoate, and methyl 2-hydroxy-3-methylbutyrate, the latter which was only found in cooked tomatillo.

Carotenoid Related Volatiles

Carotenoid-derived terpene compounds comprise a third class of volatiles identified in tomatillo flavor. The oxidative decomposition of carotenoids, particularly lycopene and β-carotene, has previously been shown to lead to the formation of terpene and terpene-like compounds in tomato flavor (*6*). Unique terpenes identified in tomatillo include α-terpinolene, terpinen-4-ol, myrtenol, and cadinene. Identifications of other terpene-derived tomatillo volatiles previously reported in tomato include 6-methyl-5-hepten-2-one, geranylacetone, β-ionone, and β-damascenone. β-Damascenone has previously been shown to arise from thermal pH 4 hydrolysis of tomato glycosides (*9*), and this finding is supported by its increased level in cooked vs. fresh tomatillo (Table III). Other terpenes including limonene, linalool, neral, and geranial were also identified. The biopathway derivation of these compounds has been previously established for tomato (*9*). Cumic alcohol (*para*-(*iso*-propyl)benzyl alcohol) was identified only in cooked tomatillo, and exhibits a caraway-like odor. It is presumed to be produced via hydroxylation and aromatization of α-terpinene.

Lignin Related Volatiles

A series of aromatic compounds including benzaldehyde, benzyl alcohol, methyl benzoate, ethyl benzoate, and methyl salicylate were identified in fresh and cooked tomatillo at different ratios. Benzaldehyde and benzyl alcohol very probably result from glycoside hydrolysis in the fresh tomatillo. Buttery (*18*) has observed benzaldehyde in the products of enzymatic hydrolysis of tomato glycosides.

Amino Acid Derived Volatiles

In comparison to the volatile profile of fresh tomato, notably absent in volatiles identified from tomatillo were the amino acid-derived volatiles including *iso*-butylthiazole, nitrophenylethane, and phenylacetonitrile. These compounds play a considerable role in tomato flavor and are also present in the mature green stage of tomato development (9).

Determination of Primary Odorants in Tomatillo

Many of the identified compounds in Table III have minimal significance for re-creation of tomatillo flavor. From this list, a more pinpointed subset of key aroma compounds was desired for a better understanding of key odorants which contribute to tomatillo. GC-olfactometry techniques such as aroma extract dilution analysis (AEDA) (*19*) or Charm analysis (*20*) are often applied to elucidate the key odorants in flavor isolates. However, these techniques are somewhat time-consuming and require repetitive analyses. Our concern was that relative concentrations of flavorants could change during the time interval required for multiple analyses.

One approach which has been successfully used to determine key aroma compounds in other fruits and vegetables utilizes the calculation of odor units (U_o). (*10*) This technique requires access to published threshold values for specific compounds, or for newly identified compounds, the determination of individual odor thresholds in water. Many of the volatiles identified in tomatillo have been previously reported in other flavor studies (*10,21*).

The odor unit is defined as the ratio of a flavor compound's concentration divided by its odor threshold:

$$U_o = \frac{\text{Compound Concentration}}{\text{Odor Threshold}}$$

The logarithm of the odor threshold (log U_o) is calculated to represent changes in concentration which are significant for olfactory discrimination. Odor activity responses in biological organisms require order-of-magnitude changes in concentration for significant differences to be produced. Consequently, logarithmic functions more significantly represent meaningful sensory differences. Aroma unit values >1 are indicative of compounds present at a concentration that greatly exceeds their thresholds, and therefore are likely to contribute significant flavor impact.

A comparison of flavor significant volatiles in fresh and cooked tomatillo is shown in Table IV. Of interest is that the aldehydic components (*Z*)-3-hexenal, nonanal, hexanal, decanal, and (*E*)-2-hexenal are present in tomatillo at order-of-magnitude higher concentrations than in tomato (*5*), and thus should contribute considerable influence to its green, leafy, fatty-soapy flavor character. During the cooking process, the aldehydic components (*Z*)-3-hexenal, nonanal, hexanal, (*E*)-2-hexenal and the alcohols 1-hexanol and (*Z*)-3-hexen-1-ol are considerably reduced. Alternatively, two aroma-significant components benzaldehyde and methyl 2-hydroxy-3-methylbutyrate, were enhanced after heat treatment.

Sensory Significant Volatile Comparison of Fresh and Cooked Tomatillo

Ten compounds with the greatest odor unit (U_o) values in fresh tomatillos are represented as a bar graph of their log values in Figure 2 (light bars). The most significant volatile is (*Z*)-3-hexenal, followed by β-ionone and (*E,E*)-2,4-decadienal. The (*E,Z*) isomer of decadienal, which contributes an oily, fatty character, also appears in the priority ranking. In comparison, a graph of the seven most significant volatiles in cooked tomatillo is also displayed in Figure 2 (dark bars). During the cooking process, the flavor contribution of the aldehydic components is slightly diminished, with the exception of (*E*)-2-octenal. β-Damascenone is an important contributor for both.

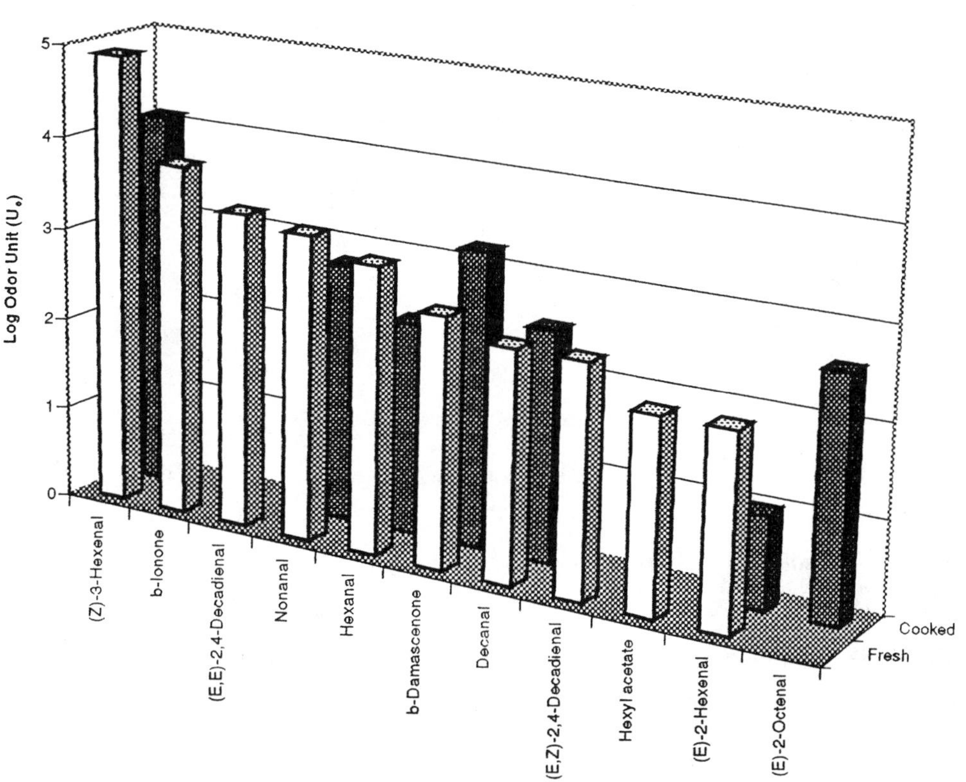

Figure 2. Log odor unit values for significant contributors to fresh (light bars) and cooked (dark bars) tomatillo aroma.

Table IV. Major Flavor-Significant Volatiles in Fresh and Cooked Tomatillo

Compound	Odor Thresh. *(ppb in H_2O)*[a]	Fresh Conc. *(ppb)*	Fresh Log Odor U[b]	Cooked Conc. *(ppb)*	Cooked Log Odor U[b]
(*Z*)-3-Hexenal	0.25	21,000	4.9	3,300	4.1
β-Ionone	0.007	49	3.8		
(*E,E*)-2,4-Decadienal	0.07	160	3.4		
Nonanal	1	2,000	3.3	700	2.8
Hexanal	4.5	5,400	3.1	820	2.3
β-Damascenone	0.002	1	2.7	3	3.2
Decanal	1	320	2.5	300	2.5
(*E,Z*)-2,4-Decadienal	0.07	22	2.5		
Hexyl acetate	2	250	2.1		
(*E*)-2-Hexenal	17	2,000	2.1	160	1.0
1-Hexanol	500	26,000	1.7	7,300	1.2
(*Z*)-3-Hexen-1-ol	70	2,700	1.6	800	1.1
Benzaldehyde	350	540	0.2	2,200	0.8
(*E*)-2-Octenal	1	tr.		32	2.6

[a] Values obtained from Refs. 10 and 21.

[b] Logarithm of compound Concentration divided by its Odor Threshold.

Conclusion

While the volatile composition of vegetable products have been studied over the past two decades, the great majority of these works have presented the qualitative and quantitative composition, but not the relevance of the identifications.

The objective of this study was to isolate and identify the volatile and non-volatile flavor components in fresh and cooked tomatillos with a goal of applying quantitative techniques which enable identification and classification of the most potent volatile compounds with sensory signficance. The dynamic headspace islolation technique was chosen to minimize potential artifact formation which could result from using thermally-intensive methods.

The volatile flavor profile of fresh tomatillos was found to be dominated by aldehydes and alcohols including (*Z*)-3-hexenal, (*E,E*)-2,4-decadienal, nonanal, hexanal, hexanol, and (Z)-3-hexen-1-ol, which provide a dominant "green flavor" impact due to their high log Odor Unit values. While other classes of volatile compounds were identified as similar to those reported in fresh tomato, the tomatillo aroma profile did not contain characteristic key tomato volatiles such as *iso*-butylthiazole, nitrophenylethane, or phenylacetonitrile. Compounds unique to tomatillo flavor included hydroxy esters, aromatic esters, 8- to 12-carbon aldehydes, decanoic acid and terpenes. The volatile profile of cooked tomatillos exhibited a 7-fold reduction of aldehyde levels compared to the fresh vegetable.

Acknowledgments

The authors graciously thank Dr. Marlene A. Stanford for proposing the project idea, and for assistance in preparing cooked tomatillo samples with carefully controlled heating profiles.

References

1. Cantwell, M.; Flores-Minutti, J.; Trejo-Gonzalez, A., *Scientia Horticulturae* **1992**, *50*, 59.
2. Tindall, H. D. *Vegetables in the Tropics*, Avi Pub., Westport, CN, 1983, pp. 359, 378.
3. Uhl, S. *Food Technology*, **1996**, *7*, 79.
4. Juttelstad, A. *Food Formulating*, **1997**, *3*, 46.
5. McGorrin, R. J.; Gimelfarb, L. In: *Food Flavors: Formation, Analysis, and Packaging Influences. Proceedings of the 9th International Flavor Conference*, Contis, E. T.; Ho, C.-T.; Mussinan, C. J.; Parliment, T. H.; Shahidi, F.; Spanier A. M., Eds., Developments in Food Science Vol. 40, Elsevier, New York, NY, , 1998, pp. 295-313.

6. Petro-Turza, M. *Food Rev. Int.* **1987**, *3*, 309.
7. Thakur, B. R.; Singh, R. K.; Nelson, P. E. *Food Rev. Int.*, **1996**, *12*, 375.
8. Buttery, R. G.; Teranishi, R.; Flath, R. A.; Ling, L. C. In: *Flavor Chemistry: Trends and Developments*, Teranishi, R.; Buttery, R. G.; Shahidi, F., Eds.; ACS Symposium Series No. 388, American Chemical Society, Washington, DC, 1989, pp 213-222.
9. Buttery, R. G.; Ling, L. C. In: *Bioactive Volatile Compounds from Plants*, Teranishi, R.; Buttery, R. G.; Sugisawa, H., Eds.; ACS Symposium Series No. 525, American Chemical Society, Washington, DC, 1993, pp 23-34.
10. Buttery, R. G. In: *Flavor Science: Sensory Principles and Techniques*, Acree, T. E.; Teranishi, R., Eds.; ACS Professional Reference Book, American Chemical Society, Washington, DC, 1993, pp 259-286.
11. Karmas, K.; Hartman, T. G.; Salinas, J. P.; Lech, J.; Rosen, R. T. In: *Lipids in Food Flavors*, Ho, C.-T.; Hartman, T. G., Eds.; ACS Symposium Series No. 558, American Chemical Society, Washington, DC, 1994, pp 130-143.
12. Servili, M..; Selvaggini, R.; Begliomini, A. L.; Montedoro, G. F.. In: *Food Flavors: Formation, Analysis, and Packaging Influences. Proceedings of the 9th International Flavor Conference*, Contis, E. T.; Ho, C.-T.; Mussinan, C. J.; Parliment, T. H.; Shahidi, F.; Spanier A. M., Eds., Developments in Food Science Vol. 40, Elsevier, New York, NY, 1998, pp. 315-329.
13. Ulberth, F. *J. AOAC Int.* **1994**, *77*, 1326.
14. AOAC Official Method 982.14, *Official Methods of Analysis of AOAC Intl.*, 16th ed., Arlington, VA, 1995, Chapter 32, p.30.
15. Klein, H.; Leubolt, R. J. Chromatogr. **1993**, *640*, 259.
16. Gould, W. A. *Tomato Production, Processing and Technology*, 3rd ed., CTI Publications, Inc., Baltimore, MD, 1992, p. 437.
17. Takeoka, G. In: *Flavor Chemistry: Thirty Years of Progress*, Teranishi, R; Wick, E. L.; Hornstein, I., Eds., Kluwer Academic/Plenum, New York, NY, 1999, pp. 287-304.
18. Buttery, R. G.; Takeoka, G.; Teranishi, R.; Ling, L. C. *J. Agric Food Chem.* **1990**, *38*, 2050.
19. Grosch, W. *Trends Food Sci. Technol.*, **1993**, *4*, 68.
20. Acree, T. E. In: *Flavor Science: Sensory Principles and Techniques*, Acree, T. E.; Teranishi, R., Eds.; ACS Professional Reference Book, American Chemical Society, Washington, DC, 1993, pp 1-18.
21. Maarse, H., Ed., *Volatile Compounds in Foods and Beverages*, Marcel Dekker, New York, NY, 1991

Correlation between Sensory Properties and Chemical Structures of Flavor Components

Chapter 8

Descriptors for Structure–Property Correlation Studies of Odorants

Helmut Guth, Katja Buhr, and Roberto Fritzler

Bergische Universität Gesamthochschule Wuppertal, Gauss-Strasse, 20, D–42097 Wuppertal, Germany

Odor qualities, odor threshold values and binding properties to biopolymers of odorants depend on the molecular structure. Physico-chemical descriptors for odorants can be used to correlate odor-activity and binding strength to various biopolymers. One approach to investigate structure activity relationship (SAR) between lactones was made by the measurement of lipophilicity and hydrogen bond strength. Partition coefficients in octanol/water (Log P_{oct}) and cyclohexane/water (Log P_{cyc}) solvent systems are standard parameters for expression of lipophilicity and hydrogen bond strength, respectively. Molecular modelling methods can be used to calculate further descriptors such as the shape of the molecules, the molecular electrostatic potential (MEP), the polarisability and the free energy of solvation. Conformations of lactones were generated by Monte-Carlo- and molecular dynamics-simulations. It is shown that a self-organized neural network such as the one introduced by Kohonen can be used to characterize the shape of the molecular surfaces. Superpositions of lactones and generation of two-dimensional maps by neural network were used for comparison of compounds. Surface properties such as MEP were mapped on the Kohonen neurons and used for calculation of similarity indices of odorants. Large differences in odor qualities, odor threshold values and binding properties to proteins observed for the various compounds, clearly demonstrated that structure-activity was significantly influenced by the molecular structure and electronic properties.

Introduction

Odor qualities, odor threshold values and binding affinities to biopolymers are important properties of odorants. While binding of odorants to food matrix may result in a decrease of flavor intensity of an aroma compound (*1*), similar interactions mechanisms are supposed to be responsible for binding

of a molecule to the olfactory receptor (*2*). Nose receptor cells, which may be also regarded as biopolymers, are G-protein-coupled receptors, including seven transmembrane protein domains spaning the lipid layer (*3*). In general binding properties to biopolymers depend on the three dimensional structure and on the electronic properties of odorants.

The eight stereoisomers of wine lactone (*4*), which show extreme differences in odor threshold values, are quite impressive examples for stereochemical structure effecting molecular properties.

Therefore one important task in flavor chemistry is to determine binding strength to proteins and odor activities of odorants. The compounds of interest have to be synthesized and the odor thresholds and qualities have to be determined by gas chromatography-olfactometry (GC-O) (*5*) and binding affinities by e.g. equilibrium dialyses (*6*) or exclusion size chromatography (*7*). Because of this time consuming procedure it is a goal in flavor chemistry to develop a method that allows one to estimate odor thresholds and binding affinities to proteins of unknown compounds.

To solve this task for structure-activity relationships the following attempts were made and published in the literature:

Greenberg (*8*) used the so called log P value which is an indicator for hydrophobic properties to predict odor threshold values of unknown compounds. Edwards and Jurs (*9*) correlated odor intensities with structural properties of odorants by the number of double bonds, charge on the most negative atom and by a polarity parameter of a molecule. The authors obtained for a set of 58 compounds a correlation coefficient of 0.87. The model was not able to differentiate between enantiomers. Turin (*10*) used a vibrational theory for the prediction of odour qualities of compounds. This method was also not able to differentiate between enantiomers. Shvets and Dimoglo (*11*) used an electron-topological method (ETM). The model includes the calculation of electronic parameters and a distance matrix. Chastrette and Rallet (*12*) looked for structure-minty odour quality relationships by calculation of the root mean square distance (RMSD) factor in the class of terpenes. Liu and Duan (*13*) found specific structural characteristics of musk odorant molecules.

In most studies mentioned above the prediction of odor qualities of compounds was possible, but no correlations were made regarding structure odor intensities of compounds.

Systematic investigations of structure odor relationships including measurement of lipophilicity, threshold values and computational chemistry techniques as molecular modelling and neural networks were made by Guth et al.

(*14*). By the application of neural networks to various lactones the authors found strong correlations between the three dimensional structure of a molecule and its odor threshold value.

The goal of our research work was to develop a model for structure activity relationships in the class of monocyclic lactones which enables us to predict binding affinities to proteins.

Experimental

Protein Binding

Binding experiments have been carried out at constant concentrations of bovine serum albumin (BSA) and ß-lactoglobulin (ß-LG, variant B) (4.40×10^{-4} Mol/L) and variable concentrations of various γ- and δ-lactones (1.0×10^{-2} - 2.0×10^{-6} Mol/L) in a phosphate buffer solution (0.066 Mol/L, pH 7.0) by ultracentrifugation (Centricon 10: 10000 MW cut-off, Millipore, Eschborn, Germany). The protein-odorant mixture (2 ml) was incubated for 30 min at room temperature and then centrifugated (3400 rpm) for 5 min. The obtained filtrate (~ 0.2 ml) was discarded and the protein-odorant buffer solution centrifugated for further 45 min. The obtained filtrate (~ 0.7 ml) and the retentate (~ 1.1 ml) were used for the determination of the odorant concentration in protein and non-protein fraction by capillary gas chromatography/mass spectrometry (HRGC/MS). Quantitation experiments of lactones were performed as follows: to protein-odorant solution (0.5 ml), which contains bound and free odorant, and to non-protein solution (0.5 ml), which contains free odorant, an internal standard was added; for γ-lactones the corresponding δ-lactones and for δ-lactones the corresponding γ-lactones were used. After addition of internal standard, the fractions were stirred for 15 min and then extracted twice with pentane (2x2 ml). Combined extracts were dried over anhydrous sodium sulfate and then concentrated by micro distillation (*15*). HRGC/MS analyses were performed by means of a HP 5890 gas chromatograph connected to a HP 5971 mass spectrometer (Hewlett Packard, Böblingen, Germany) operating in the EI-mode. HRGC separation of odorants was performed on a DB-FFAP capillary (J&W Scientific, Fisons, Mainz, Germany). The samples were applied by the on-column injection technique at 35^0C. After 2 min, the temperature of the oven was raised at 40^0C/min to 60^0C and held for 1 min isothermally, then raised at 8^0C/min to the final temperature of 240^0C. For quantitation of γ-lactones and δ-lactones by mass spectrometry the molecular ions m/z 85 and m/z 99 were monitored and used for calculations of odorant concentrations. The following mass spectrometer correction factors for γ-lactones and δ-lactones, respectively, were used for calculation of analyte concentration: γ-heptalactone (0.055), γ-nonalactone

(0.438), γ-undecalactone (0.609), δ-heptalactone (18.078), δ-nonalactone (2.282) and δ-undecalactone (1.641).

The concentration of odorant bound by the protein results from the difference in lactone amount in protein and non-protein fraction. For each binding isotherm 5 to 10 different odorant concentrations, depending on the solubility of the analyte, were applied whereas protein concentration remained constant; binding constants have been calculated from the binding isotherms by non-linear regression (software package: Origin 5.0™, Microcal Software, Northampton, USA) for a one-site binding model.

Partition coefficient (log P)

Partition coefficients (log P values) of odorants were determined by the shaking-flask method (*16*) in solvent mixtures of octanol/water (Log P_{Oct}) and cyclohexane/water (Log P_{Cyc}). To a mixture of water (4 ml) and organic solvent (3.9 ml) a solution of the odorant in cyclohexane and octanol (0.1 ml, 5 -10 µg), respectively, was added and stirred for 30 min. After centrifugation (5 min, 3400 rpm) the organic layer was separated from the water layer and an internal standard (corresponding γ-lactone and δ-lactone, respectively) was added to both fractions. Quantitation of odorant in the water- and organic fraction was performed by HRGC/MS analysis as described above for protein binding experiments.

Molecular Modelling

The following molecular modelling methods were applied to obtain energy minimized molecular structures of odorants and protein-odorant complexes:

For the generation of energy minimized structures the software package HyperChem 5.0 (Hypercube, Gainesville, Florida, USA) was used. Conformations of lactones were generated by Monte-Carlo- (MC) and molecular dynamics- (MD) simulations. On the energy minimized structures (MM+) a Connolly surface was computed using a shell 2.0 times the van der Waals radii of the atoms, with a density of 10 points per $Å^2$. The molecular electrostatic potential (MEP) was computed on the surface by application of the semi-empirical method AM1 (software:WinMopac v.2.0, Mopac 97, Fujitsu, Chiba, Japan) according to the Orozco-Luque model (*17*) and by the split-valence-basis-set 6-31G* (HyperChem 5.0). The data visualization was achieved by the software Iris Explorer 3.5 (NAG, Oxford, UK).

Sculpt 2.5 (Interactive Simulations, San Diego, USA) was used for protein-odorant docking experiments. The ß-lactoglobulin-retinol complex was downloaded via Internet access from the Brookhaven Protein Data Bank (PDB).

After construction of the odorant-protein complex in the hydrophobic binding pocket of the protein, the resulting structure was energy-minimized for van der Waals- and electrostatic interactions by the sculpt minimizer (*18*). Various orientations of the odorant in the hydrophobic binding pocket of ß-LG were tested and energy-minimized.

Neural Network

A self-organized neural network (software package: Neural Network 2.0, SPSS Science, Erkrath, Germany) such as the one introduced by Kohonen (*19*) were used to characterize the shape of the molecular surfaces (*20*), expressed as x,y,z-coordinates.

The parameters of the network were: input data not normalized, Euclidean distance error response, dimension of the network 19x19 nodes,

learning rate 0.05, decay 0.1, neighborhood size 5 nodes, neighborhood decay rate 1.0, weights (3x19x19=1083) were set randomly, random seed 100.

Superposition of lactones with template molecule and generation of two-dimensional maps by neural network were used for comparison of lactones. Surface properties such as MEP were mapped onto the Kohonen neurons and used for visual inspection of the odorants.

Results and Discussion

Protein Binding Studies

A selection of 6 monocyclic lactones (γ- and δ-heptalactone, γ- and δ-nonalactone, γ- and δ-undecalactone) were used in our studies. Binding constants, resulting from binding isotherms (examples in Figure 1) of lactones to BSA and ß-LG are summarized in Table I.

Binding affinities of lactones to BSA and ß-LG were investigated by a ultracentrifugation technique. Experimental findings (*21*) have shown that the binding equilibria are not disturbed by this procedure if the filtered volume does not exceed 40% of the initially introduced volume.

Data for odorant-protein binding are presented in the format of bound odorant plotted against the logarithm of free odorant (Figure 1). In this case the graph has some characteristic features: an inflection point appears at half-maximum binding (*22*). Position of the inflection point will be the half number of receptor sites (n). From the graphs in Figure 1 it is obvious that a clear inflection point was observed for the binding isotherm of γ-heptalactone to ß-LG. For the other compounds presented in Figure 1 and listed in Table 1 we are not able to obtain an inflection point because of their low solubility in the water system (cf. Table I). The highest solubility was found for γ-heptalactone, followed by δ-heptalactone and δ-nonalactone (Table I). In the class of γ-lactones the solubility decreases by a factor

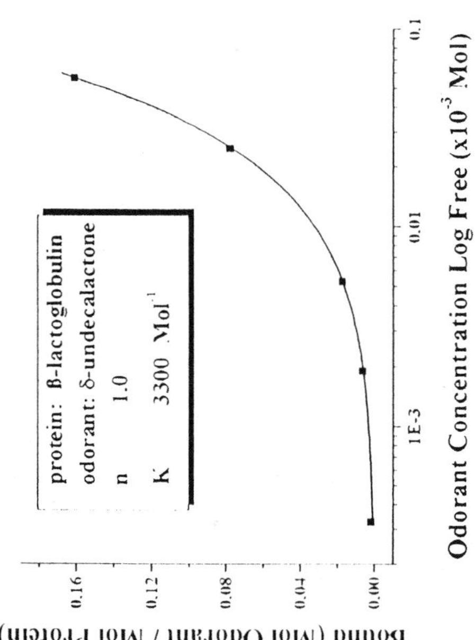
protein: ß-lactoglobulin
odorant: δ-undecalactone
n 1.0
K 3300 Mol⁻¹
Bound (Mol Odorant / Mol Protein)
0.16
0.12
0.08
0.04
0.00
1E-3
0.01
0.1
Odorant Concentration Log Free (x10⁻³ Mol)

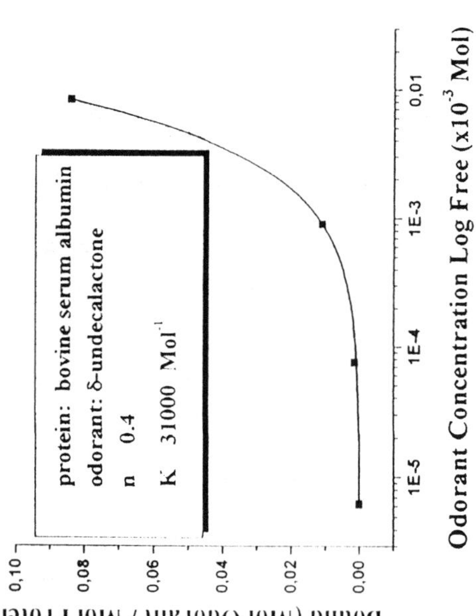
protein: bovine serum albumin
odorant: δ-undecalactone
n 0.4
K 31000 Mol⁻¹
Bound (Mol Odorant / Mol Protein)
0,10
0,08
0,06
0,04
0,02
0,00
1E-5
1E-4
1E-3
0,01
Odorant Concentration Log Free (x10⁻³ Mol)

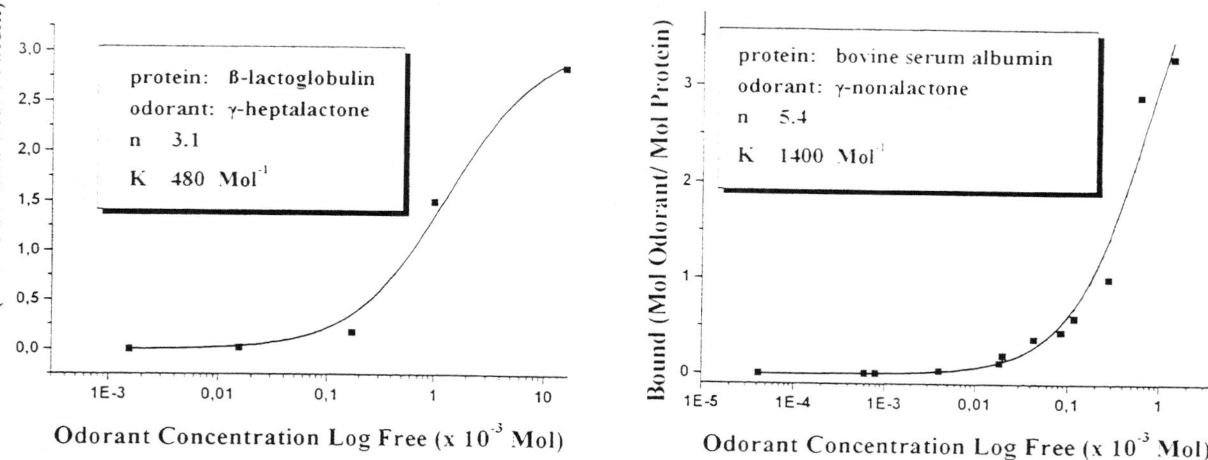

Figure 1. Binding isotherms of various lactones to ß-lactoglobulin (ß-LG) and bovine serum albumin (BSA).

Table I. Association constants (K) and receptor sites (n) for the binding of various lactones to bovine serum albumin (BSA) and ß-lactoglobulin (ß-LG) and their solubilities

Compound[a]	K (Mol^{-1}) BSA	n BSA	K (Mol^{-1}) ß-LG	n ß-LG	Solubility[b] (mmol/L)
γ-Undecalactone	2.8×10^4	2.1	1.1×10^4	0.5	0.87
γ-Nonalactone	1.4×10^3	5.4	0.74×10^3	2.1	12.5
γ-Heptalactone	3.9×10^2	1.2	4.8×10^2	3.1	211
δ- Undecalactone	3.1×10^4	0.4	3.3×10^3	1.0	2.9
δ- Nonalactone	7.4×10^2	3.0	$< 10^2$	-	33.9
δ- Heptalactone	$< 10^2$	-	$< 10^2$	-	61

[a] No differences were found for the binding affinities of the corresponding R- and S-enantiomers to ß-LG and BSA.
[b] Maximum solubility (mmol/L) of lactones in phosphate buffer solution.

of 16.9 from γ-heptalactone to γ-nonalactone and by a factor of 14.4 from γ-nonalactone to γ-undecalactone. In contrast the series of δ-lactones showed a decrease of the solubility only by a factor of 1.8 between δ- heptalactone and δ-nonalactone and a factor of 11.7 comparing δ- nonalactone and δ- undecalactone.

The calculation of binding sites and binding constants were achieved by statistical evaluation. For a detailed discussion on the background of the calculation of binding constants see reference (*22*) and (*23*).

No differences were found for the binding affinities of the corresponding R- and S-enantiomers to ß-LG and BSA. High association constants were found for γ- (2.8×10^4 Mol^{-1}) and δ- undecalactone (3.1×10^4 Mol^{-1}) for the binding to BSA. The binding strength of these two compounds to ß-LG are lower compared to BSA.

In the series of γ-lactones the binding affinity to BSA decreases by a factor of 20 and 72, respectively, from γ-undecalactone to γ-nonalactone and γ-heptalactone. In general the binding affinities of lactones to the proteins decreased with decreasing carbon units of lactones.

Lipophilicity and Hydrogen Bond Strength

One approach to investigate structure activity relationships is the measurement of lipophilicity and hydrogen bond strength as indicators for enthalpic

(ΔH) and entropic (ΔS) promotion of binding to a receptor protein (*24*). Possible ligand binding sites at the protein may be located where hydrogen bonds can be formed and hydrophobic pockets can be occupied by lipophilic groups of the ligand. It was assumed that the last mentioned process has a significant contribution to protein-ligand binding. Indicator variables for lipophilicity are log P values which are calculated from partition coefficients in octanol/water solvent systems. The log P_{Oct} values for the investigated lactones were determined and are summarized in Table II.

In the series of γ- and δ-lactones Log P values (cf. Table II) increased with increasing carbon chain length, corresponding to higher lipophilicity of compounds. Log P values of γ-lactones are higher compared to the corresponding δ-lactones.

Comparison of lactone-binding constant (K) to BSA and ß-LG (cf. Table I), respectively, to their corresponding Log P values by linear regression analyses indicate strong correlation (R=0.975) for Log P_{Oct} and binding constants of lactones to BSA (Figure 2).

Table II. Partition coefficients (Log P_{Oct} and Log P_{cyc}), calculated Δ Log $P_{Oct - Cyc}$, minimum molecular electrostatic potential (MEP_{Min}) and partial charge of various lactones

Compound	Log P_{Oct}	Log P_{Cyc}	Δ Log $P_{Oct - Cyc}$	MEP_{Min} [a]	Partial charge [b]
γ-Undecalactone	3.30	2.52	0.78	-65.2	-0.543
γ-Nonalactone	1.95	1.22	0.73	-65.2	-0.543
γ-Heptalactone	0.65	-0.10	0.75	-65.2	-0.543
δ- Undecalactone	2.93	2.10	0.83	-69.8	-0.568
δ- Nonalactone	1.54	0.76	0.78	-69.8	-0.568
δ- Heptalactone	0.27	-0.51	0.78	-69.8	-0.568

[a] Molecular electrostatic potential was calculated according to Ford and Wang (*25*) with the semi-empirical method AM1.

[b] Sum of the partial charges on the two oxygen atoms in the lactone ring (Mopac 97, AM1).

Similar results were found by Sostmann and Guichard (26) for a series of aldehydes, ketones and unsaturated alcohols (C_4-C_{10}) for the binding to ß-LG; increasing carbon chain length increased the binding affinities to ß-LG by a constant factor. But the authors did not correlate binding affinities with lipophilicity of odorants.

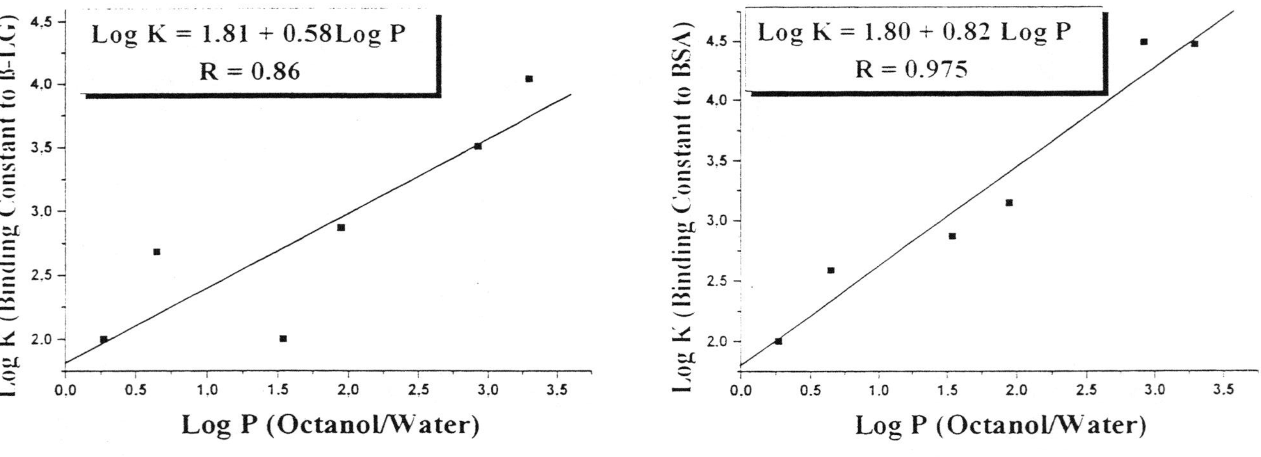

Figure 2. Correlation of lactone Log P values with binding affinities to bovine serum albumin (BSA) and ß-lactoglobulin (ß-LG).

According to the lower correlation coefficient (R=0.86) for the binding of lactones to ß-LG (cf. Figure II) it can be concluded that besides lipophilicity other factors are important for the protein-binding strength.

One further aspect in protein-ligand binding is the enthalpic contribution (ΔH) to the free binding enthalpy (ΔG). One type of reaction is the formation of hydrogen bonds of ligand to the receptor. A molecule has to slip off its hydrogen shell in order to proceed from water to cyclohexane. While this is not necessary for passover from water to octanol, the difference in Log P values for both systems may be regarded as a measure for hydrogen bond strength (*27*). Calculated Δ Log P values are summarized in Table II. From the data in Table II it is obvious that the γ-lactones show slightly lower Δ Log P values compared to δ-lactones, corresponding to weaker hydrogen bond formation strength. These results are in agreement with the calculated molecular electrostatic potential minimum (MEP_{Min}) values and the partial charges (cf. Table II) of γ- and δ-lactones: MEP_{Min} values and partial charges of γ-lactones are lower compared to the corresponding δ-lactones. This fact could be an explanation for the formation of weaker hydrogen bonds of γ-lactones due to the less electrophilic potential compared to δ-lactones. According to the above mentioned results δ-lactones should have a stronger binding affinity to receptor molecules if hydrogen bonding plays an important role. But in contrast we found higher binding constants of γ-lactones compared to δ-lactones. From these observations we conclude that further descriptors of the compounds as for example the three dimensional structure are important for their binding affinity.

Molecular Modelling and Neural Networks

In previous studies (*14*) on structure-odor relationships of various mono- and bicyclic lactones it could be shown that molecular modelling techniques in combination with neural networks are a powerful tool to estimate odor threshold values of unknown compounds.

The numeric information (x,y,z-coordinates and MEP at odorant surface point), on which the displayed graphical representation is based, was subjected to further similarity studies by neural network (*14*). Therefore the three dimensional data (x,y,z-coordinates) were projected onto a two dimensional Kohonen Map and afterwards the electrostatic potential of the data points were projected to each neuron (Figure 3).

The three dimensional structures of odorants including the MEP were reduced to two dimensions without loss of information (cf. Figure 3). In order to compare the different surfaces a single template approach (*14,28*) has been applied: after training of the network with the reference molecules (R- and S-γ-

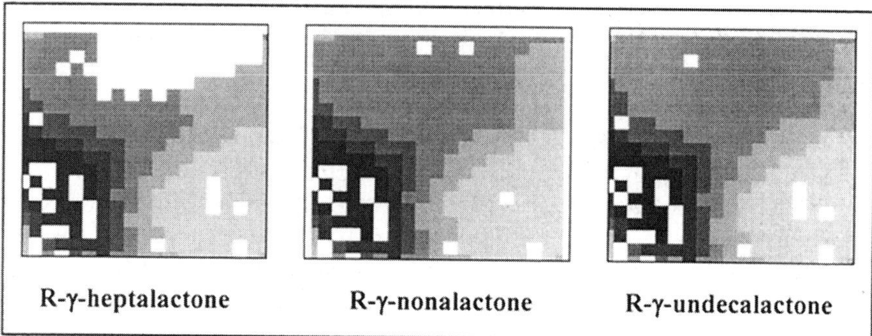

Figure 3. Molecular electrostatic potential (MEP) projected onto the Kohonen Maps of various γ-lactones. Colour coding: black= (-22.5 kcal/Mol) - (-5.0 kcal/Mol), dark grey= (-5.0 kcal/Mol) - (+ 0.5 kcal/Mol), light grey= (+0.5 kcal/Mol) - (+ 12.0 kcal/Mol), white= not occupied neurons.

undecalactone), which are supposed to have optimum properties (in our case high binding affinity to ß-LG), the surface data of the remaining molecules are projected onto the same networks: for the R-lactone enantiomers the template molecule R-γ-undecalactone was used and for the series of S-configurated lactones the template molecule S-γ-undecalactone was used. The resulting Kohonen Maps featuring the MEP may be compared visually or by statistical methods.

The Kohonen Maps in Figure 3 indicate large differences between the compounds R-γ-undecalactone and R-γ-heptalactone. Differences in side chain length of R-γ-undecalactone compared to R-γ-heptalactone yielded white areas in Kohonen Map of R-γ-heptalactone due to unoccupied neurons. The same differences were obtained for the S-configured lactones by comparison with template molecule S-γ-undecalactone (data not shown).

A more objective representation can be achieved by displaying the difference spectrum compared to the template molecule as shown in Figure 4 for R-γ-heptalactone and R-γ-nonalactone. Large differences were observed for R-γ-heptalactone which also shows lower binding affinity to ß-LG. Specific maps with large differences compared to the reference compound can be used to locate areas in the space of a molecule, which are responsible for higher binding affinities. This is achieved by backprojection of Kohonen map neurons onto the molecular structure of the odorant (achieved by the software package IRIS EXPLORER). The observed differences correlate very well with the observed binding affinities of lactones to ß-LG. The backprojection of Kohonen map neurons of R-γ-undecalactone which are not occupied by e.g. R-γ-heptalactone (cf. Figure 3 and 4) onto the molecular structure of R-γ-heptalactone indicate that the location in space and the length of the hydrophobic alkyl chain, fused to the lactone ring, is mainly responsible for stronger binding affinity of R-γ-undecalactone compared to R-γ-heptalactone.

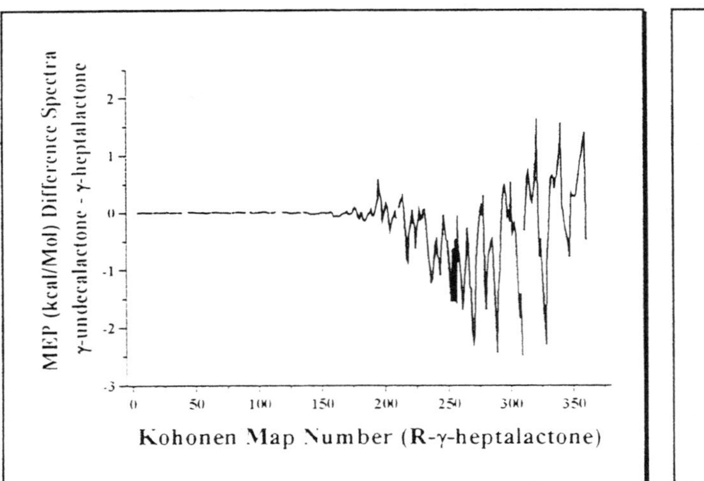

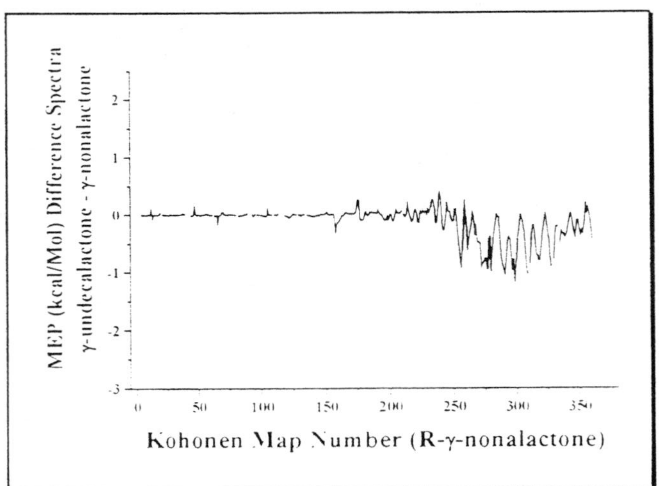

Figure 4. Molecular electrostatic potential (MEP) difference spectra obtained by substracting MEP´s in each neuron (361 neurons) of R-γ-heptalactone and R-γ-nonalactone, respectively, from the corresponding MEP neuron of R-γ-undecalactone (=template molecule).

Docking Experiments

The application of receptor-ligand docking experiments by molecular modelling to predict binding affinities of unknown compounds to a receptor molecule is possible if the structure of the receptor molecule is known. In our case the molecular structure of ß-LG (variant B) as well as the retinol binding protein (RBP) complex was taken from the Brookhaven Protein Databank (PDB) via Internet access. The RBP complex was energy minimized with the software package Sculpt (Figure 5). After that RBP and retinol were replaced by ß-LG and by different lactones, respectively. The resulting protein-ligand-complexes were optimized by the Sculpt minimizer for van der Waals and electrostatic

interactions in the hydrophobic binding pocket of ß-LG (Figure 5). During this calculation the protein structure was fixed and the lactone compound was allowed to move. The differences in resulting energies of energy-minimized ß-LG-lactone complexes and retinol binding protein (RBP) complex agree with the observed differences in binding constants (Figure 5). Binding constant of RBP-retinol complex was taken from the literature (*29*). Based on our observation by molecular modelling and on the amino acid similarity and structural homology of ß-LG to RBP, one can suggest that ß-LG binds the lactones in the central pocket. From literature two retinol binding sites were proposed for ß-LG (*30*), one in the center of the calyx and one in a surface groove. Our molecular modelling experiments suggested that lactones bound in the central pocket (Figure 5) of ß-LG.

The energy difference for ß-LG-R-γ-nonalactone- (E=-25.0 kcal/Mol) and ß-LG-R-γ-undecalactone (E=-28.1 kcal/Mol) complex was calculated to 3.1 kcal/Mol, corresponding to 2.1 orders of magnitude higher binding affinity for R-γ-undecalactone compared to R-γ-nonalactone. From experiments (cf. Table I and Figure 5) the difference in binding affinity of R-γ-undecalactone and R-γ-nonalactone corresponds to 1.48 orders of magnitude. Thus the determined binding affinities are in good agreement with the theoretical values found by molecular modelling experiments.

Acknowledgments

Financial support of the Deutsche Forschungsgemeinschaft (DFG, project 515/1-1) and the Heinrich Hertz-Stiftung is gratefully acknowledged.

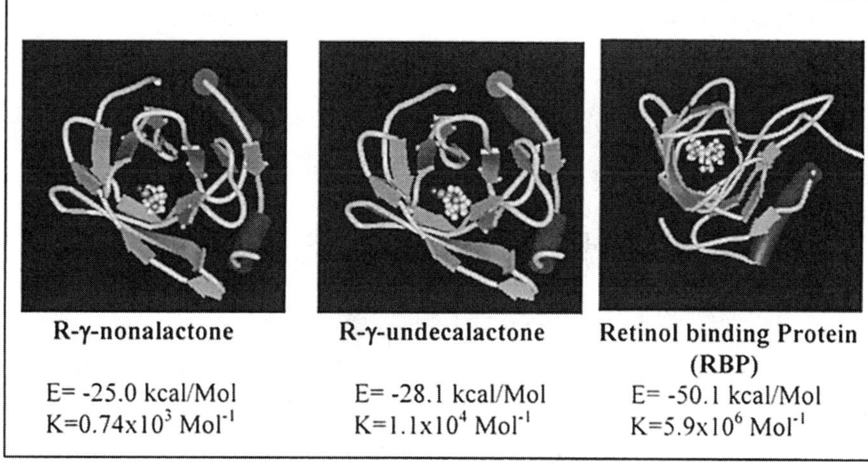

Figure 5. Schematic representation of energy-minimized structures of retinol binding protein (RBP) and ß-lactoglobulin (ß-LG) bound with various lactones.

Literature Cited

1. O'Neill, T. E.; Kinsella, J. E., *J. Agric. Food Chem.* **1987**, 35, 770-774.
2. Snyder, S. H.; Sklar, P. B.; Pevsner, J., *J. Biol. Chem.* **1988**, 263, 13971-13974.
3. Breer, H.; Raming, K.; Krieger, J., *Biochim. Biophys. Acta* **1994**, 1224, 277-287.
4. Guth, H., *Helv. Chim. Acta* **1996**, 79, 1559-1571.
5. Ullrich, F.; Grosch, W. , *Z. Lebensm. Unters. Forsch.* **1987**, 184, 277-282.
6. Fares, K.; Landy, P.; Guilard, R.; Voilley, *J. Dairy Sci.* **1998**, 81, 82-91.
7. Guichard, E.; Etievant, P., *Nahrung* **1998**, 42, 376-379.
8. Greenberg, M.J.; *J. Agric. Food Chem.* **1979**, 27, 347-352.
9. Edwards, P.A.; Jurs, P., *Chem. Senses* **1989**, 14, 281-291.
10. Turin, L., *Chem. Senses* **1996**, 21, 773-79.
11. Shvets, N.M.; Dimoglo, A.S., *Nahrung* **1998,** 42, 364-370.
12. Chastrette, M.; Rallet, E., *Flavour Fragrance J.* **1998**, 13, 5-18.
13. Liu, J.; Duan, G., *J. Mol. Struct.* **1998**, 432, 97-103.
14. Guth, H.; Buhr, K.; Fritzler, R., In *Flavour Science*, Schieberle, P.; Engel, K.-H., Ed.; 9th Weurman Flavour Research Symposium, Freising, Germany, 1999, in press.
15. Bemelmans, J. M. H. In Progress in flavor research, Land, D. G.,Nursten, H. E., Ed.; Applied Science, London, 1979, pp 79-88.

16. Leo, A.; Hansch, C.; Elkins, D., *Chem. Rev.* **1971,** 71, 525-616.
17. Luque, F. J.; Orozco, M., *Chem. Phys. Lett.* **1990,** 168, 269–275.
18. Surles, M.; Richardson, J.; Richardson, D.; Brooks, F. *Protein Science* **1994**, 3, 198.
19. Kohonen, T. In *Self-organizing maps*; Huang, T.S., Kohonen, T., Schröder, M.R., Ed.; Springer, Berlin, Germany, 1995.
20. Anzali, S.; Barnickel, G.; Krug, M.; Sadowski, J.; Wagener, M.; Gasteiger, J.; Polanski, J., *J. Comput.-Aided Mol. Design* **1996,** 10, 521-534.
21. Whitlam, J. B.; Brown, K. F., *J. Pharm. Sci.* **1981,** 70, 146-150.
22. Klotz, I. M., *Science* **1982**, 217, 1247-1249.
23. Lindup, W. E., In *Progress in Drug Metabolism*, Bridges, J. W., Chasseaud, L. F., Gibson, G. G., Ed.; 1987, Vol. 10, Taylor & Francis Ltd, pp 141-185.
24. Böhm, H.-J., In *Structure-based ligand design*, Mannhold, R., Kubinyi, H., Timmerman, H., Ed.; 1998, Wiley-VCH, Weinheim, Germany, pp 129-142.
25. Ford, G.P.; Wang, B., *J. Comp. Chem.* **1993**, 14, 1101-1111.
26. Sostmann, K.; Guichard, E., *Food Chem.* **1998**, 62, 509-513.
27. Seiler, P., Eur. *J. Med. Chem.* **1974**, 9, 473-479.
28. Anzali, S.; Barnickel, G.; Krug, M.; Sadowski, J.; Wagener, M.; Gasteiger, J.; Polanski, J., *J. Comput.-Aided Mol. Design* **1996**, 10, 521-534.
29. Cogan, U.; Kopelman,M.; Mokady, S.; Shinitzky, M., *Eur. J. Biochem.* **1976**, 65, 71-78.
30. Batt, A.C.; Brady, J.; Sawyer, L., *Trends Food Sci. Technol.* **1994**, 5, 261-265.

Chapter 9

Influence of the Chain Length on the Aroma Properties of Homologous Epoxy–Aldehydes, Ketones, and Alcohols

P. Schieberle and A. Buettner

Deutsche Forschunganstalt fuer Lebensmittelchemie, Lichtenbergstrasse 4, 85748 Garching, Germany

Homologous series of epoxyalkanals, epoxyalkenals, alka-1,5-dien-3-ones, alka-1,5-dien-3-ols, alk-1-en-3-ones and alk-1-en-3-ols were synthesized and their aroma properties (odor quality; odor threshold) were determined by means of GC/Olfactometry. The results showed odor thresholds in the same order of magnitude (2.6-15.4 ng/L in air) for the 2,3-epoxyalkanals from C-5 to C-12. Contrary, in the series of 4,5-epoxy-(E)-2-alkenals a clear minimum was found for the C-10 homologue (0.001 ng/L in air). 1-Nonen-3-one, exhibiting a mushroom-like aroma, was found to be the most potent among the series of 1-alken-3-ones (C-5 to C-13), whereas the geranium-like smelling 1,(Z)-5-heptadien-3-one showed the lowest threshold (0.0006 ng/L) among the 1,(Z)-5-alkadien-3-ones. Reduction of the ketones into the respective alcohols increased the odor thresholds significantly.

Introduction

The peroxidation of unsaturated fatty acids is one of the key reaction pathways leading to either desired or undesired aromas during food processing and storage. Depending on the position and geometry of the double bonds in the fatty acid moiety and, also, on the different mechanisms of hydroperoxide formation and degradation (cf. review in (*1*)), a tremendous number of volatile degradation products is predictable and many of them have already been identified (cf. review in (*1*)). It is, however, well known, that the chemical structure has much influence on the aroma attributes, in particular, the odor quality and the odor threshold. For instance, the shift of the double bond from the 4 to the 6 position and an inversion of the geometry of the double bond changes the odor quality of nonadienals from deep-fat fried (2,4-(E,E)-isomer) to cucumber-like (2,6-(E,Z)-isomer; Table 1). The introduction of a further double bond into 1-octen-3-one changes the aroma from mushroom- to geranium-like in 1,(Z)-5-octadien-3-one and, also, reduces the odor threshold (Table 1).

Table 1. Aroma attributes of selected odorants known to be formed by fatty acid degradation (cf. review in *1*).

Odorant	Odor quality	Odor threshold[a] (μg/L in water)
(E,E)-2,4-Nonadienal	deep-fat fried	0.07
(E,Z)-2,6-Nonadienal	cucumber-like	0.01
1-Octen-3-one	mushroom-like	0.005
1,(Z)-5-Octadien-3-one	geranium-like	0.001

[a] Data from (*2*)

This fact has inspired researchers already in the early seventies to systematically study the influence of, e.g., chain length, number of double bonds or their geometry on the aroma properties of, in particular aldehydes. Some of the results are reviewed in Tables 2 and 3.

Saturated aldehydes show different odor qualities for the C-6 up to the C-8 compound, whereas octanal, nonanal and decanal elicit a nearly identical aroma quality. The odor threshold, however, is lowest for the C-10 aldehyde (Table 2). The introduction of a trans-double bond in the 2-position completely changes the odor quality compared to the respective saturated aldehyde. However, only in case of the pair nonanal/(E)-2-nonenal the odor threshold was significantly changed (Table 2).

Table 2. Aroma attributes of some aldehydes.

Aldehyde	Odor quality	Odor threshold[a] (ng/L air)
Hexanal	green, grassy	40
Heptanal	fatty	250
Octanal	soapy, citrus-like	11
Nonanal	soapy, citrus-like	5
Decanal	citrus-like, soapy	1
(E)-2-Hexenal	bitter-almond-like	125
(E)-2-Heptenal	bitter-almond, fatty	53
(E)-2-Octenal	fatty, nutty	47
(E)-2-Nonenal	tallowy, fatty	0.1
(E)-2-Decenal	tallowy	2.7

[a] Data summarized in (*2*) also available at www.odor-thresholds.de

The influence of the position of the double bond and its geometry is exemplified for nonenal in Table 3. If a trans double bond is moved continuously from the 2 to the 7 position, a sudden change in the odor quality from linoleic acid hardening- to melon-like is observed between the (E)-6- and (E)-7-nonenal. By far the lowest odor threshold is, however, shown by (E)-6-nonenal.

Table 3. Influence of the position and geometry on the odor thresholds and odor qualities of nonenals. Data from (*3*).

Odorant	Odor quality	Odor threshold (mg/kg in paraffin)
(E)-2-Nonenal	starch-glue	3.5
(E)-3-Nonenal	n.d.	n.d.
(E)-4-Nonenal	linoleic acid hardening	2.0
(E)-5-Nonenal	linoleic acid hardening	0.45
(E)-6-Nonenal	linoleic acid hardening	0.005
(E)-7-Nonenal	melon-like	1.0
(Z)-2-Nonenal	n.d.	n.d.
(Z)-3-Nonenal	green, cucumber	0.25
(Z)-4-Nonenal	green, cucumber	0.08
(Z)-5-Nonenal	green, melon-like	0.25
(Z)-6-Nonenal	melon-like	0.04
(Z)-7-Nonenal	melon-like	0.4

n.d.: not determined

In case of the respective (Z)-nonenals, the odor quality significantly changes from cucumber-like to melon-like between the (Z)-4- and the (Z)-5-nonenal. However, although also the (Z)-6-nonenal showed the lowest odor threshold, this minimum was not as pronounced as found for the (E)-isomers.

Although some 1-alken- and 1,5-alkadien-3-ones and their corresponding alcohols as well as epoxyalkenals are known as important food aroma compounds, no systematic studies on the influence of the chain length on the aroma properties have yet been reported. Such data would, however, facilitate the identification of, e.g., yet unknown food aroma compounds in particular those having extremely low odor thresholds and would, also, increase the knowledge on compounds fitting best into human odorant receptors. So, the issue of this work was to fill up this gap by using synthesized reference compounds and GC/Olfactometry as the bioassay to determine the aroma attributes.

Experimental Procedures

Syntheses

1-Alken-3-ols and 1-alken-3-ones were synthesized by reacting the respective two carbon atoms shorter, saturated aldehyde in a Grignard-type reaction with vinyl magnesium bromide to yield the 1-alken-3-ol which was subsequently oxidized into the ketone using Dess-Martin-periodinane (Fig. 1). The structure of the compounds was confirmed by mass spectrometry and their retention indices were determined on three GC-columns of different polarity (*4*).

1,(Z)-5-Alkadien-3-ols and 1,(Z)-5-alkadien-3-ones were prepared by a stereo-specific reduction of 3-alkyne-1-ols into the corresponding (Z)-3-alken-1-ols and then oxidized by means of Dess-Martin-periodinane. The aldehydes were subsequently reacted with vinyl magnesium bromide to yield the 1,(Z)-5-alkadien-3-ols. Oxidation with Dess-Martin-periodinane then afforded the corresponding ketone (Fig. 2). The mass spectral data and the retention indices are reported in (*4*).

The epoxyalkenals and the epoxyalkanals were synthesized as previously reported (*5, 6*).

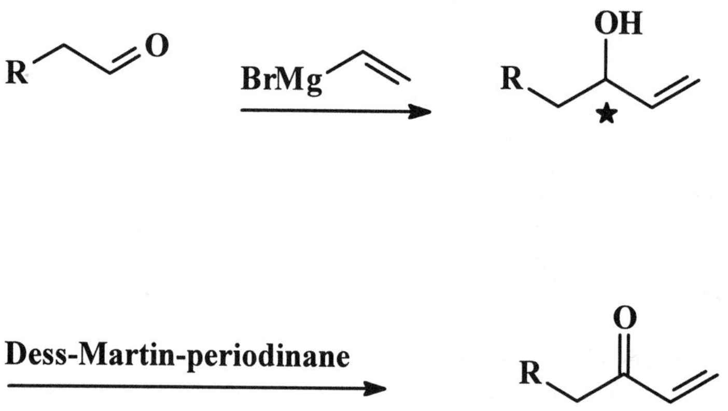

R O BrMg R OH

Dess-Martin-periodinane R O

Figure 1. Synthetic route used for the preparation of 1-alkene-3-ols and –3-ones.

High-Resolution Gas Chromatography/Olfactometry (HRGC/O); determination of odor thresholds

HRGC was performed with a Type 8000 gas chromatograph (Fisons Instruments, Mainz, Germany) by using an FFAP-column (30 m x 0.32 mm fused silica capillary, free fatty acid phase FFAP, 0.25 µm; Chrompack, Mühlheim, Germany). The samples were applied by the on column injection technique at 35°C. After 2 min, the temperature of the oven was raised at 40°C/min to 60°C, held for 2 min isothermally, then raised at 6°C/min to 180°C, and finally raised at 10°C/min to 230°C and held for 10 min. The flow rate of the carrier gas helium was 2.5 ml/min. At the end of the capillary, the effluent was split 1:1 (by vol) into an FID and a sniffing port using two deactivated but uncoated fused silica capillaries (50 cm x 0.32 mm). The FID and the sniffing port were held at 220°C.

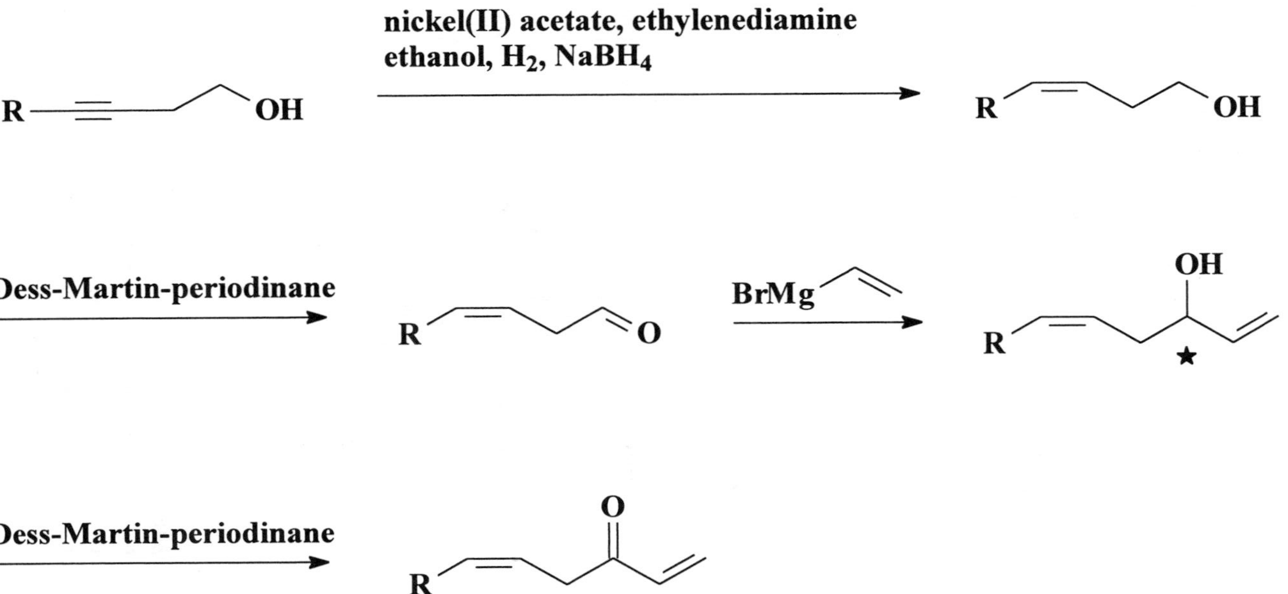

Figure 2. Synthetic route used for the preparation of 1,(Z)-5-alkadien-3-ones and 1,(Z)-5-alkadien-3-ols.

(E)-2-Decenal, for which an odor threshold in air of 2.7 ng/L has been reported (*7*) was used as the reference in the determination of odor thresholds by GC/Olfactometry. The method runs as follows: 1 mg of (E)-2-decenal and 1 mg of each of the odorants under investigation was dissolved in 10 mL of diethyl ether. 0.5 µL of this mixture were then analyzed by GC/O representing 50 ng of each compound, and of the reference (E)-2-decenal. Serial dilutions of the mixture (1:1; 1:2; 1:4 ...) were sniffed, thereby determining the last dilution in which either the odorant or the reference compound could be sniffed. Using this approach, for the reference compound a dilution factor of 32 was determined corresponding to 0.77 ng of (E)-2-decenal. If the odorant under investigation gave the same dilution factor, the odor threshold in air was calculated to be the same as that of (E)-2-decenal, namely 2.7 ng/L. If a dilution of 64 was determined for the odorant the threshold is calculated as 1.35 ng/L and so on.

Results and Discussion

Epoxyaldehydes

A homologous series (C-5 to C-12) of tr-2,3-epoxy alkanals was synthesized and their odor qualities and odor thresholds were determined by GC/O. The odor qualities were found to be green and grassy, malty for C-5 to C-7, whereas the larger homologues exhibited citrus-like, soapy aromas (*4*). Interestingly, there were no significant differences between the odor thresholds (Fig. 3), which varied from about 3 ng/L (epoxy octanal) to about 1 ng/L (epoxy dodecanal). In the homologous series of the 4,5-epoxy-(E)-2-alkenals (C-7 to C-12) a metallic odor quality prevailed (*4*). However, the odor thresholds dramatically decreased with decreasing the chain length and a minimum was observed for the 4,5-epoxy-(E)-2-decenal.

It is interesting to note that among the important food odorants only the tr-2,3-epoxy octanal (*7, 8*) has been reported, although it might be speculated that also other (E)-2-alkenals known as food constituents (e.g. (E)-2-hexenal) might react with peroxides to yield the respective epoxy alkanal. Among the series of 4,5-epoxy-(E)-2-alkenals, however, three homologues have been reported as food volatiles (Table 4).

1-Alken-3-ones; 1-Alken-3-ols

In a further series of experiments, 1-alken-3-ones (C-5 to C-13) were synthesized and their aroma attributes were determined. There was a large variation in the aroma qualities. E.g., the 1-penten-3-one showed a pungent odor, whereas the 1-hexen- and 1-hepten-3-one were described as metallic, vegetable-like (*4*). 1-Octen- and 1-nonen-3-one clearly exhibited mushroom-like aromas, whereas the larger homologues were characterized by herb-, citrus-like odor qualities (*4*). The odor thresholds were found to slightly decrease from C-5 to C-9 (Fig. 4), but were increased with further increasing the chain length to C-13.

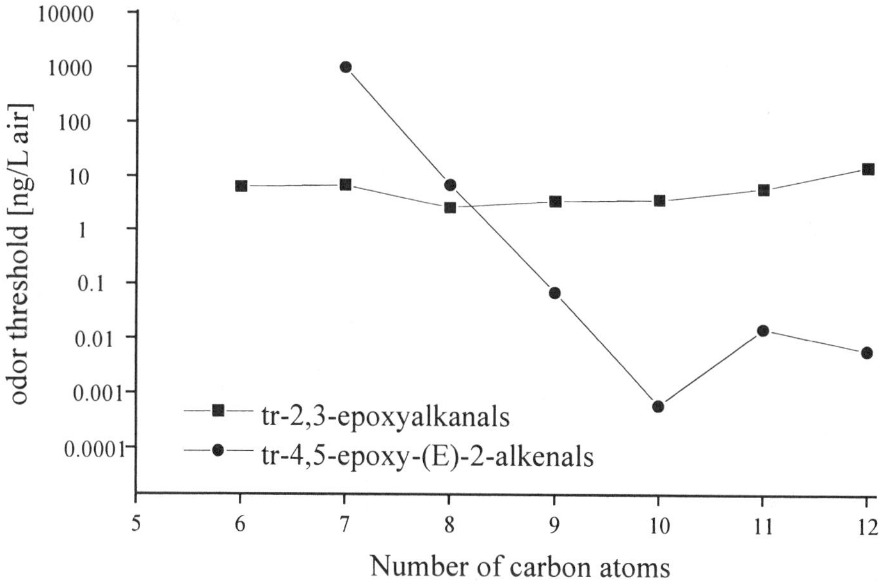

Figure 3. Influence of the chain length on the odor thresholds of homologous tr-2,3-epoxyalkanals and tr-4,5-epoxy-(E)-2-alkenals.

Table 4. Epoxyaldehydes previously reported as important odorants in food.

Odorant	Previously identified in
tr-4,5-Epoxy-(E)-hept-2-enal	butterfat (4)
tr-4,5-Epoxy-(E)-non-2-enal	soya bean oil, rapeseed oil (7)
tr-4,5-Epoxy-(E)-dec-2-enal	soya bean oil, rapeseed oil (7), wheat bread crumb (5)

The reduction of the oxo function did not have a significant effect on the odor qualities (*4*). However, the odor thresholds were increased by factors of about 8 (C-12) to 500 (C-7), when compared with the respective ketone (Fig. 4).

It is interesting to note that six of the 1-alken-3-ones (C-5 to C-9 and C-12) have already been identified in foods (Table 5). However, the C-10 and C-11 ketones are reported here for the first time. 1-Hepten-3-one was very recently identified as key odorant in grapefruit juice (*17*).

Among the series of 1-alken-3-ols, 1-penten-3-ol up to 1-nonen-3-ol have previously been reported as food volatiles (Table 6). The higher homologues, however, have not yet been reported as food aroma compounds.

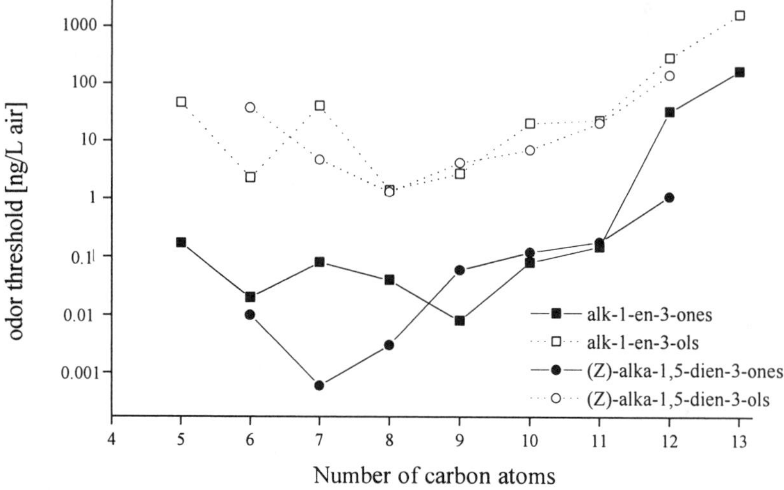

Figure 4. Influence of the chain length on the odor thresholds of homologous (Z)-alka-1,5-dien-3-ones, (Z)-alka-1,5-dien-3-ols, alk-1-en-3-ones and alk-1-en-3-ols.

Table 5. Alk-1-en-3-ones previously reported as important odorants in food.

Odorant	Previously identified in
Pent-1-en-3-one	banana (11), tea (12), tomato (13)
Hex-1-en-3-one	artichoke (14), honey (15), dill (16)
Hept-1-en-3-one	grapefruit juice (17)
Oct-1-en-3-one	artichoke (14), mushrooms (18), soya bean oil (10)
Non-1-en-3-one	raspberry (19), yogurt (20), sheep meat (21)
Dodec-1-en-3-one	white bread (22)

Table 6. Alk-1-en-3-ols previously reported as important odorants in food.

Odorant	Previously identified in
Pent-1-en-3-ol	banana (11, 23), dill herb (24)
Hex-1-en-3-ol	banana (23), dill herb (24), tea (12)
Hept-1-en-3-ol	banana (11, 23), dill herb (24)
Oct-1-en-3-ol	banana (11, 23), fish (25, 26)
Non-1-en-3-ol	banana (11, 23)

1,(Z)-5-Alkadien-3-ones; 1,(Z)-5-Alkadien-3-ols

In the series of homologous 1,(Z)-5-alkadien-3-ones, a geranium-like aroma was found for the C-6 up to the C-8 homologue. However, this odor quality was changed to herb-, mushroom-like for the higher homologues (*4*). A clear minimum in the odor thresholds was found for 1,(Z)-5-heptadien-3-one (0.0006 ng/L) which was by a factor of about 2000 lower than that of the 1,(Z)-5-dodecadien-3-one (1.1 ng/L) (Fig. 4). With the exception of the C-9 to the C-11 homologue, the odor thresholds were lower compared to that of the respective mono unsaturated ketones (Fig. 4).

As found for the 1-alken-3-ones, also the dienones were clearly increased in their odor thresholds by a reduction into the corresponding alcohol. In this homologous series, the lowest threshold was found for the 1,(Z)-5-octadien-3-ol (1.3 ng/L).

It is interesting to note that among the dienones only the 1,(Z)-5-octadien-3-one is a known contributor to food aromas, e.g. mushrooms (*18*) or soya bean oil (*10*). Among the corresponding alcohols, 1,(Z)-5-octadien-3-ol was characterized in mushrooms (*18*) and fish (*25, 26*) and 1,(Z)-5-undecadien-3-ol in fish (*25, 26*).

References

1. Grosch, W. In: *Autoxidation of unsaturated lipids*; Chan, H.W.-S.; Ed.; Academic Press Inc., London 1987, pp. 95.
2. Rychlik, M.; Schieberle, P.; Grosch, W. *Compilation of odor thresholds, odor qualities and retention indices of key food odorants*; Deutsche Forschungsanstalt für Lebensmittelchemie, 1998.
3. Meijboom, P.W.; Jongenotter, G.J. *J. Am. Oil Chem. Soc.* **1981**, 680-682.
4. Buettner, A., Schieberle, P. *J. Agric. Food Chem.* **2000**, submitted
5. Schieberle, P.; Grosch, W. *Z. Lebensm.-Untersuch. und -Forsch.* **1991**, *192*, 130-135.
6. Guth, H.; Grosch, W. *Z. Lebensm.-Untersuch. und -Forsch.* **1993**, *196*, 22-28.
7. Ullrich, F.; Grosch, W. *Z. Lebensm. -Unters. und -Forsch.* **1987**, 184, 277.
8. Morello, M. *Food flavors: formation, analysis and packaging influences*; Contis, E.T.; Ho, C.-T.; Mussinan, C.J.; Parliment, T.H.; Shahidi, F.; Spanier, A.M.; Eds.; Elsevier Science B.V., 1998, pp. 415-422.
9. Swoboda, P.A.T.; Peers, K.E. *J. Sci. Food Agric.* **1978**, *29*, 803-807.
10. Guth, H.; Grosch, W. *Lebensmittel-Wiss. Technol.* **1990**, *23*, 59-65.
11. Tressl, R.; Drawert, F.; Heimann, W.; Emberger, R. *Z. Naturforsch. B* **1969**, *24*, 781-783.
12. Mick, W.; Schreier, P. *J. Agric. Food Chem.* **1984**, *32*, 924-929.
13. Buttery, R.G.; Teranishi, R.; Ling, C.L. *J. Agric. Food Chem.* **1987**, *35*, 540-544.
14. Buttery, R.G.; Guadagni, D.G.; Ling, L.C. *J. Agric. Food Chem.* **1978**, *26*, 791-793.
15. Blank, I.; Fischer, K.-H.; Grosch, W. *Z. Lebensm.-Untersuch. und -Forsch.* **1989**, *189*, 426-433.
16. Blank, I.; Grosch, W. *J. Food Sci.* **1991**, *56*, 63-67.

17. Buettner, A.; Schieberle, P. *J. Agric. Food Chem.* **1999**, *47*, 5189-5193.
18. Tressl, R.; Bahri, D.; Engel, K.-H. *J. Agric. Food Chem.* **1982**, *30*, 89-93.
19. Roberts, D.; Acree, T. *J. Agric. Food Chem.* **1996**, *44*, 3319-3325.
20. Ott, A.; Fay, L.B.; Chaintreau, A.C. *J. Agric. Food Chem.* **1997**, 45, 850-858.
21. Rota, V.; Schieberle, P. *J. Agric. Food Chem.* **2000**, submitted.
22. Obretenov, T.; Hadjieva, P. *Z. Lebensm.-Untersuch. und -Forsch.* **1977**, *165*, 195-199.
23. Tressl, R.; Drawert, F.; Heimann, W.; Emberger, R. *Z. Lebensm.-Untersuch. und -Forsch.* **1970**, *142*, 313-321.
24. Schreier, P.; Drawert, F.; Heindze, I. *Lebensmittel-Wiss. Technol.* **1981**, *14*, 150-152.
25. Yajima, I.; Nakamura, M.; Sakakibara, H.; Ide, J.; Yanai, T.; Hayashi, K. *Agric. Biol. Chem.* **1983**, *47*, 1755-1760.
26. Sakakibara, H.; Ide, J.; Yanai, T.; Yajima, I.; Hayashi, K. *Agric. Biol. Chem.* **1990**, *54*, 9-16.

Chapter 10

Flavor Chemistry of Peppermint Oil (*Mentha piperita* L.)

Matthias Güntert[1], Gerhard Krammer[2], Stefan Lambrecht[2], Horst Sommer[2], Horst Surburg[3], and Peter Werkhoff[2]

[1]Flavor Division, R&D, Haarmann & Reimer, 300 North Street, Teterboro, NJ 07608
[2]Flavor Division and [3]Fragrance Division, R&D, Haarmann & Reimer GmbH, P.O. Box 1247, D–37603 Holzminden, Germany

The flavoring of foods and oral care products with mint is economically of growing importance. The flavorings consist mainly of blends of peppermint (*Mentha piperita*), spearmint (*Mentha spicata*), and cornmint oils (*Mentha arvensis*), along with menthol and other flavor materials. In addition, more and more nature-identical (synthetic) mint oil compositions are used.
The most valuable oil is *Mentha piperita* which is mainly grown in the United States but also in smaller amounts in France, Italy, China, and India, in various varieties. The five primary American mint oil producing regions are the Midwest Range (Indiana, Wisconsin, and Michigan), Willamette (West Oregon), Madras (East Oregon), Kennewick (Washington) and Idaho.
In our study a Willamette peppermint oil was extensively investigated by GC, GC-MS and GC-olfactometry for its flavor chemistry. The isolation and identification of so far unknown trace components was achieved by micro-preparative chromatographic methods and subsequent application of ^{1}H-NMR and ^{13}C-NMR spectroscopy. Some new flavor compounds could be identified which have not been described in peppermint oil or even in natural products before.

The use of natural mint oils and synthetic mint flavorings for the flavoring of oral care, chewing gum and confectionery products is of growing importance. The oil obtained by steam distillation of the fresh overground portion of the herb *Mentha x piperita* L. is the only one which may be called peppermint oil. It may be rectified by distillation, but is neither partially nor wholly dementholized. The estimated production of peppermint oil (*Mentha piperita* L.) in the United States counted in 1995 for approximately 4,700 metric tons.

The composition of mint oils has been extensively studied over the years (*1-8*). A particular aspect has always been the adulteration of the more valuable peppermint oil with the less costly cornmint oil and the analytical proof thereof. Therefore many articles dealt with analytical possibilities tracing such adulterations. On the other hand, not very much has been published so far about the sensory contribution of individual flavor components to the odor and flavor profile of peppermint oil. With this regard the paper from Benn (*7*) is of particular interest since the relatively new aroma extract dilution analysis (AEDA) was used for the first time to characterize odorants in Yakima peppermint oil and Chinese cornmint oil.

Experimental Procedures

Sample Preparation of the Peppermint Oil

The peppermint oil used in this study was an unrectified oil from the Willamette Valley in West Oregon. The composition of its main components is shown in Table 1 and in Figure 1. It was distilled into several fractions. Two of the fractions with germacrene D and viridiflorol being the main components were further pursued. These two sesquiterpenes themselves are important flavor compounds of the peppermint oil. But besides the two sesquiterpenes there must be other sensorically interesting flavor compounds since the fractions were sensorically evaluated as very interesting. GC and GC-MS as well as GC-olfactometry(GC-O) were applied to all samples.

Twenty grams of the germacrene D fraction were column chromatographed over 400 grams of silica gel 60. The non-polar hydrocarbons (68%) were removed by flashing with n-hexane while the polar components (28%) were eluted with ethyl acetate. The resulting polar fraction was sensorically evaluated as being very interesting. It was described as being "typical peppermint, herbaceous, green, hay-like." The polar fraction was subjected to GC and GC-MS as well as GC-olfactometry (GC-O).

A second twenty grams of the germacrene D fraction was further separated into about 40 subfractions by column chromatography over 400 grams silica gel 60. The liquid phase consisted of hexane with increasing amounts of ethyl acetate. The most interesting notes were found in the medium polar fractions. GC and GC-MS as well as GC-olfactometry (GC-O) were applied to the most interesting samples.

In cases where unknown components with interesting odors were detected, preparative capillary GC on these fractions was applied in order to enrich and isolate these trace components for ^{1}H-NMR experiments.

Table I: Main Flavor Components of an Unrectified Willamette Peppermint Oil (*Mentha piperita L.*)

Flavor Compound	rel. GC%
bicyclogermacrene	0.34
β-bourbonene	0.33
δ-cadinene	0.08
carvone	0.30
β-caryophyllene	1.80
β-caryophyllene oxide	0.03
1,8-cineole	4.80
α-copaene	0.04
p-cymene	0.13
β-elemene	0.14
(6*E*)-β-farnesene	0.16
germacrene D	1.70
(3*Z*)-hexenol	0.04
α-humulene	0.23
γ-3,4-dehydro-7,8-dihydro-ionol	0.03
isoamyl isovalerate	0.15
isopulegol	0.21
(*Z*)-jasmone	0.04
limonene	1.70
linalool	0.23
cis-p-menth-2-en-1-ol	0.08
trans-, trans-p-menth-2-en-1-ol	0.10
mentha-1,8(10)-dien-9-yl acetate	0.03
menthofuran	2.30
menthol	44.2
iso-menthol	0.16
neo-menthol	3.00
neoiso-menthol	0.65
menthone	20.4
iso-menthone	2.90
menthyl acetate	4.30
neo-menthyl acetate	0.16
neoiso-menthyl acetate	0.21
mintsulfide	0.03
myrcene	0.22
(*E*)-β-ocimene	0.08
(*Z*)-β-ocimene	0.25
1-octen-3-ol	0.14
3-octanol	0.26
3-octyl acetate	0.02
α-pinene	0.82
β-pinene	0.90
piperitone	0.63
pulegone	1.20
sabinene	0.43
cis-sabinene hydrate	0.11
trans-sabinene hydrate	0.83
spathulenol	0.03
α-terpinene	0.31
γ-terpinene	0.50
1-terpinen-4-ol	1.00
α-terpineol	0.35
δ-terpineol	0.14
terpinolene	0.17
thymol	0.07
1,(3*E*,5*Z*)-undecatriene	0.02
viridiflorol	0.20
Total	> 99.5%

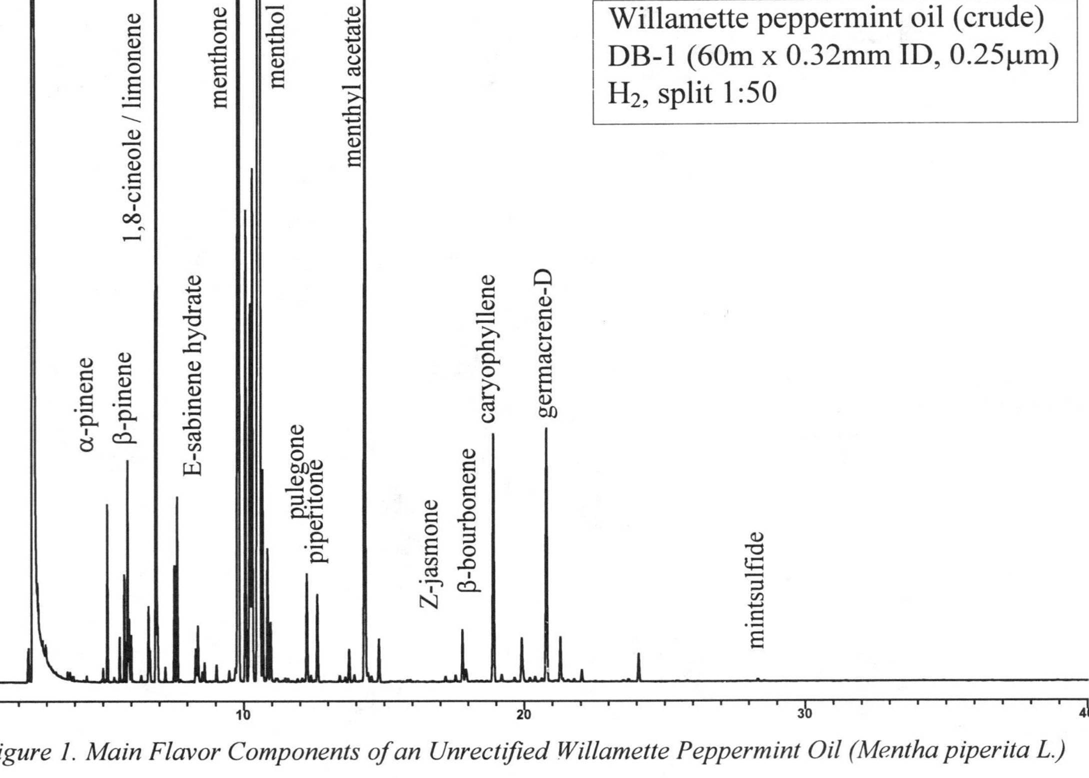

Figure 1. Main Flavor Components of an Unrectified Willamette Peppermint Oil (Mentha piperita L.)

Instrumental Analysis

Instrumentation (capillary gas chromatography, spectroscopy), analytical and micro preparative conditions were described in previous publications (*9-11*).

Component Identification

All detected flavor components of the peppermint oil and its fractions were identified by comparison of their mass spectra and retention indices (against homologous series of n-alkanes) with those of a reference standard. Most of the reference standards were characterized in the Central Analytical Services spectral libraries of H&R by a set of ^{1}H-NMR, ^{13}C-NMR, mass and infrared spectra, as well as two retention indices run on DB-1 and DB-WAX capillaries (J&W, California). The synthesis and spectra of the unknown and newly described components is shown below.

Synthesis and Spectra of the New Components

The isomers of 8-ocimenyl acetate were prepared from an (*E,Z*)-mixture of β-ocimene via the (*E,Z*)-6-chloro-α-ocimenes: 109 grams of (0.80 mol) ocimene (E/Z ratio 2:1) in 500 mL methyl-tert.-butylether were chlorinated with 100 grams (0.74 mol) of sulfuryl chloride in the presence of 85 grams sodium carbonate at room temperature for 3 h. After usual work-up, distillation led to a fraction of about 90 grams with a boiling point of 72°C and 1 mbar containing a mixture of (*E,Z*)-6-chloro-α-ocimenes in a purity (GC) of about 75%.

The chloro ocimenes were converted into the 8-ocimenyl acetates by reaction with sodium acetate at 100°C in DMF in the presence of catalytical amounts of sodium iodide. The isomers of the obtained mixture of 8-ocimenyl acetates were isolated in pure form by micro preparative capillary gas chromatography. Subsequent spectral analysis led to the result that the mixture consisted of about 30-35% of (*Z*3,*Z*6)-8-ocimenyl acetate 7, about 30-35% of (*E*3,*Z*6)-8-ocimenyl acetate 9, about 10-15% of (*Z*3,*E*6)-8-ocimenyl acetate and about 10-15% of (*E*3,*E*6)-8-ocimenyl - acetate (Patent DE 19,748,774; 5 November 1997, to Haarmann & GmbH, inventors H. Surburg, H. Sommer, S. Lambrecht, P. Wörner, M. Güntert, G. Kindel, V. Koppe).

Mass spectral data (EI, 70 eV):

(*Z*3,*Z*6)-8-ocimenyl acetate (MW 194) - 43 (100), 119 (60.4), 134 (31.9), 41 (28.8), 91 (25.4), 93 (23.9), 55 (23.8), 79 (20.3), 39 (18.2), 92 (17.8).

(*E*3,*Z*6)-8-ocimenyl acetate (MW 194) - 43 (100), 119 (69.5), 134 (35.7), 41 (27.7), 55 (24.3), 91 (23.5), 79 (20.2), 93 (19.0), 39 (17.0), 27 (16.9).

(*Z*3,*E*6)-8-ocimenyl acetate (MW 194) - 43 (100), 119 (45.7), 93 (36.0), 92 (29.1), 79 (24.9), 134 (23.5), 91 (22.0), 41 (21.0), 55 (19.5), 105 (17.0).

(*E*3,*E*6)-8-ocimenyl acetate (MW 194) - 43 (100), 119 (44.6), 134 (25.0), 41 (21.5), 93 (21.5), 79 (19.9), 55 (18.7), 91 (17.7), 39 (14.2), 27 (13.8).

^{1}H-NMR data (400 MHz, TMS, $CDCl_3$):
(*Z*3,*Z*6)-8-ocimenyl acetate - 1.756 ppm (3H, 1.3 Hz, q), 1.814 ppm (3H, 1.2 Hz, q), 2.073 ppm (3H, s), 2.954 ppm (2H, 7.5 Hz, t), 4.606 ppm (2H, s), 5.119 ppm (1H, 1.6 Hz, t, 10.8 Hz, d), 5.223 (1H, 17.3 Hz, d), 5.313 ppm (1H, 7.6 Hz, t), 5.379 ppm (1H, 7.5 Hz, t), 6.764 ppm (1H, 10.8 Hz, d, 17.3 Hz, d).
(*E*3,*Z*6)-8-ocimenyl acetate - 1.758 ppm (3H, mc), 1.760 ppm (3H, mc), 2.074 ppm (3H, s), 2.926 ppm (2H, 7.4 Hz, t), 4.610 ppm (2H, s), 4.950 ppm (1H, 10.7 Hz, d), 5.108 ppm (1H, 17.4 Hz, d), 5.39 ppm (1H, mc), 5.42 ppm (1H, mc), 6.351 ppm (1H, 10.7 Hz, d, 17.4 Hz, d).
(*Z*3,*E*6)-8-ocimenyl acetate - 1.689 ppm (3H, mc), 1.821 ppm (3H, 1.3 Hz, q), 2.074 (3H, s), 2.926 ppm (2H, 7.3 Hz, t), 4.458 ppm (2H, 1.0 Hz, d), 5.122 ppm (1H, 17.2 Hz, d), 5.228 ppm (1H, 17.2 Hz, d), 5.340 ppm (1H, mc), 5.450 ppm (1H, mc), 6.766 ppm (1H, 10.7 Hz, d, 17.2 Hz, d).
(*E*3,*E*6)-8-ocimenyl acetate - 1.693 ppm (3H, mc), 1.767 ppm (3H, 0.8 Hz, t, 1.3 Hz, d), 2.075 ppm (3H, s), 2.897 ppm (2H, 7.2 Hz, t), 4.461 ppm (2H, mc), 4.954 ppm (1H, 10.7 Hz, d), 5.111 ppm (1H, 17.4 Hz, d), 5.447 (1H, 7.3 Hz, t), 5.464 ppm (7.3 Hz, t), 6.362 ppm (1H, 10.7 Hz, d, 17.4 Hz, d).
^{13}C-NMR data (100 MHz, TMS, $CDCl_3$):
(*Z*3,*Z*6)-8-ocimenyl acetate - 132.90 ppm (C2), 62.77 ppm (C8), 21.39 ppm (C9), 19.46 ppm (C10).
(*E*3,*Z*6)-8-ocimenyl acetate - 141.02 ppm (C2), 63.18 ppm (C8), 21.37 ppm (C9), 11.46 ppm (C10).
(*Z*3,*E*6)-8-ocimenyl acetate - 132.93 ppm (C2), 68.04 ppm (C8), 13.74 ppm (C9), 19.49 ppm (C10).
(*E*3,*E*6)-8-ocimenyl acetate - 141.06 ppm (C2), 69.48 ppm (C8), 13.37 ppm (C9), 11.14 ppm (C10).
Retention indices:
(*Z*3,*Z*6)-8-ocimenyl acetate - 1363 (DB-1), 1829 (DB-WAX)
(*E*3,*Z*6)-8-ocimenyl acetate - 1375 (DB-1), 1855 (DB-WAX)
(*Z*3,*E*6)-8-ocimenyl acetate - 1393 (DB-1), 1892 (DB-WAX)
(*E*3,*E*6)-8-ocimenyl acetate - 1407 (DB-1), 1918 (DB-WAX)

3,6-Dimethylbenzo-[b]-furan-2(3H)-one (furamintone) 19 was synthesized by dehyrogenation of 5,6-dihydro-3,6-dimethyl-2(4H)-benzofuranone in the presence of palladium on charcoal as catalyst and mesityl oxide as hydrogen acceptor (Patent DE 19,909,980; 27 April 1998, to Haarmann & Reimer GmbH, inventors S. Lambrecht, H. Surburg, M. Güntert, V. Koppe).
Mass spectral data (EI, 70 eV):
162 (MW) - 134 (100), 91 (76.0), 162 (75.9), 133 (29.5), 119 (25.1), 39 (19.8), 105 (19.2), 77 (17.8), 65 (16.2), 51 (15.3).
^{1}H-NMR data (400 MHz, TMS, $CDCl_3$):
1.55 ppm (3H, 7.5 Hz, d), 2.38 ppm (3H, 0.8 Hz, d), 3.69 ppm (1H, 7.5 Hz, q), 6.90 ppm (1H, 0.6 Hz, d), 7.00 ppm (1H, 0.6 Hz, d, 7.3 Hz, d), 7.13 ppm (1H, 7.3 Hz, d).

^{13}C-NMR data (100 MHz, TMS, $CDCl_3$):
178.38 ppm (C2), 38.24 ppm (C3), 125.75 ppm (C3a), 123.52 ppm (C4), 124.77 ppm (C5), 139.20 ppm (C6), 111.30 ppm (C7), 153.58 ppm (C7a), 16.01 ppm (C8), 21.65 ppm (C9).
Retention indices: 1196 (DB-1), 1689 (DB-WAX)

The isomers of 8-ocimenol were obtained by saponification of the above mentioned isomeric mixture of the 8-ocimenyl acetates with KOH. The isomers of the crude 8-ocimenol mixture were separated by preparative capillary GC or HPLC.
Mass Spectral data (EI, 70 eV):
(*Z*3,*Z*6)-8-ocimenol (MW 152) - 119 (100), 79 (89.3), 91 (76.6), 84 (73.6), 41 (72.2), 93 (62.6), 77 (55.3), 55 (48.7), 39 (48.7), 134 (45.3).
(*E*3,*Z*6)-8-ocimenol (MW 152) - 119 (100), 79 (67.8), 91 (63.6), 134 (63.1), 41 (59.2), 84 (46.1), 55 (41.7), 93 (41.5), 77 (41.3), 39 (40.3).
(*Z*3,*E*6)-8-ocimenol (MW 152) - 79 (100), 93 (96.5), 94 (72.0), 41 (59.1), 91 (53.1), 92 (53.1), 77 (47.6), 43 (41.9), 39 (39.3), 55 (35.5).
(*E*3,*E*6)-8-ocimenol (MW 152) - 79 (100), 93 (86.3), 41 (58.2), 80 (55.0), 91 (49.3), 77 (45.7), 43 (42.9), 94 (42.2), 39 (38.2), 55 (36.9).
Retention indices:
(*Z*3,*Z*6)-8-ocimenol – 1250 (DB-1), 1960 (DB-WAX)
(*E*3,*Z*6)-8-ocimenol – 1264 (DB-1), 1988 (DB-WAX)
(*Z*3,*E*6)-8-ocimenol – 1268 (DB-1), 1994 (DB-WAX)
(*E*3,*E*6)-8-ocimenol – 1282 (DB-1), 2033 (DB-WAX)

Gas Chromatography-Olfactometry

GC with simultaneous FID and odor port evaluation was carried out using a Carlo Erba Type 5360 Mega Series gas chromatograph. Separations were performed using a 60 m x 0.32 mm (ID) capillary column coated with DB-1 (d_f = 1 µm) and DB-WAX (d_f = 0.5 µm) stationary phases. The flow rate of the helium carrier gas was 3 – 4 mL/min. The column effluent was split 1:15 with a glass-cap-cross (Seekamp, Achim, Germany). The temperature program used was 60°C – 220 °C at 3 °C /min. Injections were made in the split/splitless mode. The injector temperature was 220 °C and the detector temperature was 250 °C.

Determination of the Odor Threshold and Taste Threshold Values

The odor thresholds of the single flavor compounds were determined in air using a GC-O method very similar to the one described first by Ullrich et al. (*12,13*). First the split ration was set to 1:15. Second the ratio of the effluent going to the sniffing port was determined to be 3:1. Every reference compound was set up as a 0.1% solution in diethyl ether. Consequently, if 1 µL of the 0.1% solution is being injected into the GC injector one can calculate how much substance comes through at the sniffing port. The solutions were then diluted down by the factor 2^n and each dilution step was subjected to GC-O. The experiments were stopped when no odor could be detected anymore. On this way it was possible to determine the dilution factor and to calculate the odor threshold in air. In order to compare the measured values of the hitherto unknown compounds against known values some reference compounds (myrcene, 1,8-cineole, linalool, β-damascenone) with known odor thresholds in air were also measured using the same method. The sniffing experiments were done by two flavorists.

The taste thresholds were done by using a method described as Deutsche Industrienorm (German Industry Norm) DIN 10959. First solutions of the single flavor compounds in 5% sugar solution were made up. Then it was determined by a panel of 20 trained people how strong the respective compounds are by diluting down the solutions by a factor of 10^n. As soon as the range of the taste thresholds was known dilutions (2^n) of the respective solutions were made. The panel of 20 people then tasted the different solutions in a blind mode and had to determine in which solution they could not taste anything more (= taste threshold). They also had to determine in which solution they were just able to recognize the typical taste of the respective flavor compound (= recognition taste threshold).

Results and Discussion

Flavor Chemistry of Mint Oils

Mint oils consist to a large extent of monoterpenes and sesquiterpenes. Therefore it is not surprising that the flavor chemistry of mint oils is heavily influenced by the terpene metabolism. More than 98% by volume of the volatile flavor components of a peppermint oil are mono- and sesquiterpenes. Especially menthol and menthol derivatives dominate the flavor chemistry of mint oils. The main components of peppermint oils are menthol (30-50%), menthone (15-30%), menthyl acetate (2-9%), neomenthol (4-8%), 1,8-cineole (4-7%), isomenthone (2-4%), and menthofuran (1-7%). In Table 1 the composition of the main flavor components of the unrectified Willamette peppermint oil that was used for most of our research is shown.

But as this is the case with all essential oils and genuine food flavors the overall sensory impression of a complex and delicately balanced peppermint oil is not only represented by these main flavor components but by a large number of smaller and trace components.

In this work we focussed mainly on two different classes of flavor components that play an important sensory role in peppermint oils. The one is a range of acyclic monoterpene alcohols and their corresponding acetates and the other bicyclic monoterpene lactones.

Acyclic Monoterpene Hydrocarbons, Alcohols and Esters

They are mainly represented in peppermint oil by myrcene (0.22%) 1 as well as (*Z*)-ß-ocimene (0.25%) 2 and (*E*)-ß-ocimene (0.08%) 3. We were able for the first time to find traces of the oxidized structures of myrcene and the isomeric ocimenes, 2-methyl-6-methylene-(2*Z*),7-octadienol ((*Z*)-8-myrcenol) 4, 2-methyl-6-methylene-(2*Z*),7-octadienyl acetate ((*Z*)-8-myrcenyl acetate) 5, 2,6-dimethyl-(2*Z*,5*Z*),7-octatrienol ((*Z*,*Z*)-8-ocimenol) 6, 2,6-dimethyl-(2*Z*,5*Z*),7-octatrienyl acetate ((*Z*,*Z*)- 8-ocimenyl acetate) 7, 2,6-dimethyl-(2*Z*,5*E*),7-octatrienol ((*E*,*Z*)-8-ocimenol) 8, and 2,6-dimethyl-(2*Z*,5*E*),7-octatrienyl acetate ((*E*,*Z*)-8-ocimenyl acetate) 9 in the Willamette peppermint oil. Other peppermint oils are currently being analyzed for the detection of these components. The occurrence of hydroxyl and acetoxy groups in allylic position, especially at the chain terminus, is quite common in the flavor chemistry of various essential oils and

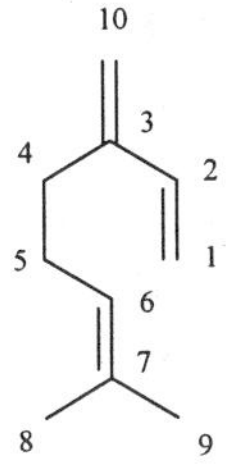

1 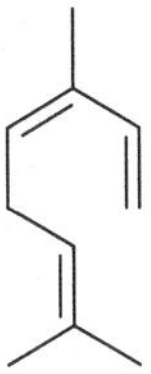 2 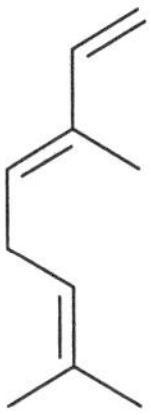 3 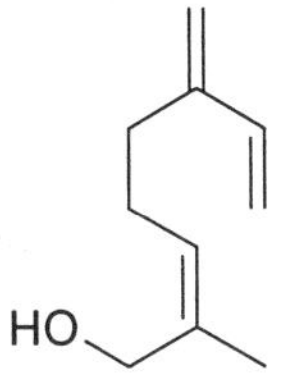 4 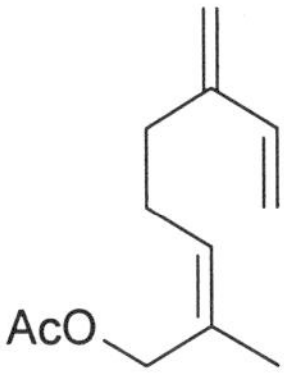 5

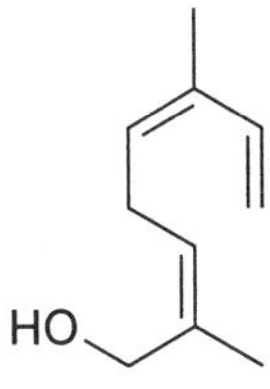

6 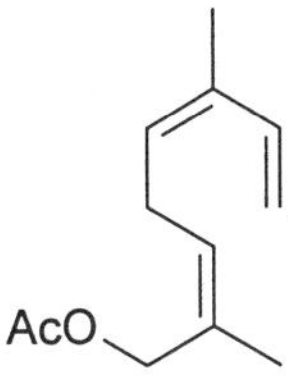 7 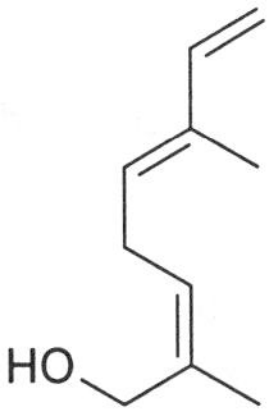 8 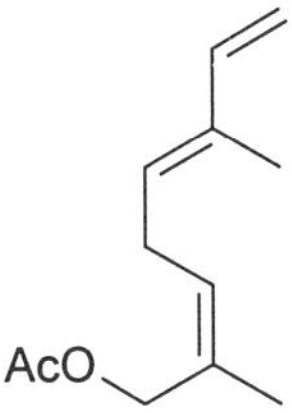 9

genuine food flavors though not so much in peppermint oil. The most well-known monoterpene alcohols in peppermint oil in allylic position are cis- and trans-2-menthen-1-ol 10 as well as cis- and trans-6,8(10)-menthadien-2-ol (carveol) 11. Also known are the monoterpene acetates in allylic position, 1,8(10)-menthadien-9-yl acetate 12 and the cis/trans-6,8(10)-menthadien-2-yl acetates (carvyl acetates) 13.

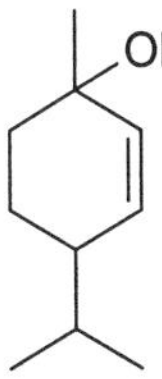

10 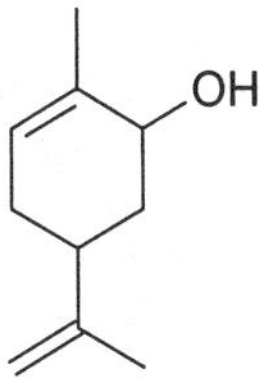 11 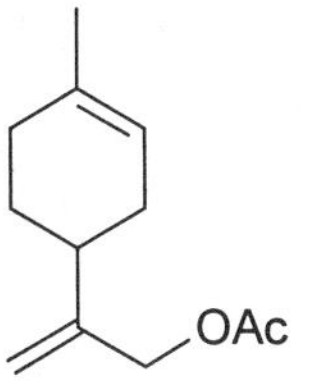 12 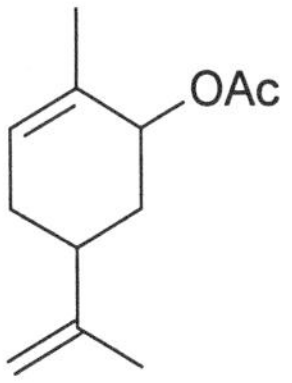 13

The monoterpene alcohol (*Z*)-8-myrcenol 4 and its acetate 5 are reported for the first time in peppermint oil. Interestingly, only the respective *Z*-configurated isomers could be identified. This allows the cautious conclusion that they are biogenerated by an enzymatic allylic oxidation of myrcene 1. This monoterpene hydrocarbon is a common component in peppermint oils. It was sensorically described as being "woody, terpeney". Its odor threshold was determined as 112.5 ng/L air.

(*Z*)-8-Myrcenol 4 and its acetate 5 are quite uncommon components in natural products. They are especially known as typical components in the essential oil of the so-called 4-thujanol chemotype of Common Thyme (*Thymus vulgaris*) (*14-16*). This particular oil contains characteristically trans-thujan-4-ol as the dominant compound (35-52%) and along with this significant amounts of (*Z*)-8-myrcenol 4 and (*Z*)-8-myrcenyl acetate 5. (*Z*)-8-Myrcenol 4 is also known from *Amomum subulatum* Roxb., an Indian large cardamom oil (*17*), and from own flavor investigations of raspberry fruits (*18*). (*Z*)-8-Myrcenyl acetate was also identified in the essential oil of lovage seeds (*Levisticum officinale* Koch.) by Toulemonde et al. (*19*). (*Z*)-8-Myrcenol 4 and (*Z*)-8-myrcenyl acetate 5 were described as strong odorants with fruity and perfume-like notes but as very unstable (*15*). (*Z*)-8-Myrcenol was sensorically described by our flavorists as "floral, lily-of-the-valley" while (*Z*)-8-myrcenyl acetate had more "herbaceous, mango-like, terpeney" aspects. The *E*-isomers were never found in essential oils or food flavors. This correlates well with our findings in peppermint oil.

The isomeric (*E*)-8-myrcenol, (*Z,E*)-8-ocimenol and (*E,E*)-8-ocimenol were tentatively characterized by Strauss et al. (*20*) in their study on the novel monoterpenes in *Vitis vinifera* grapes. But they did not find them in the grapes but only in a model system as minor transformation products from the acid hydrolysis of 2,6-dimethyl-(2*E*),7-octadiene-1,6-diol ((*E*)-8-hydroxylinalool). Consequently, neither one of the two isomeric 2,6-dimethyl-(2*Z*,5*Z*),7-octatrienol ((*Z,Z*)-8-ocimenol) 6 and 2,6-dimethyl-(2*Z*,5*E*),7-octatrienol ((*E,Z*)-8-ocimenol) 8 are known in the literature and therefore are described here for the first time. As well the respective acetates 2,6-dimethyl-(2*Z*,5*Z*),7-octatrienyl acetate ((*Z,Z*)-8-ocimenyl acetate) 7 and 2,6-dimethyl-(2*Z*,5*E*),7-octatrienyl acetate ((*E,Z*)-8-ocimenyl acetate) 9 were never described before in the literature.

The discussed monoterpene alcohols and acetates can be seen as oxidized structures of myrcene 1 as well as (*Z*)-ß-ocimene 2 and (*E*)-ß-ocimene 3 in allylic position at the chain terminus. Since there were only the (*Z*)-allylic isomers found in peppermint oil the oxidation appears to be stereospecifically controlled and therefore enzymatically governed. However this seems to be somewhat in contrast with the fact that about 3 times as much (*Z*)-ß-ocimene 2 as (*E*)-ß-ocimene 3 is present in peppermint oil. Allylic oxidations at the chain terminus are not uncommon in natural products. One of the better known examples in food is the enzymatic oxidation of linalool, nerol, geraniol, and citronellol in grape must through *Botrytis cinerea*. The monoterpene alcohols are oxidized at C-8 and form the respective 8-hydroxylinalool, 8-hydroxynerol, 8-hydroxygeraniol, and 8-hydroxycitronellol. Studies on this subject were done first by Mandery (*21*) as well as Bock et al. (*22*). Interestingly, the resulting 8-hydroxy monoterpene alcohols were mainly *E*-configured.

The above mentioned new monoterpene alcohols and acetates are listed in Table 2 with some analytical parameters and their sensory characteristics. Especially, the two isomeric acetates 2,6-dimethyl-(2*Z*,5*Z*),7-octatrienyl acetate ((*Z*,*Z*)-8-ocimenyl acetate) 7 and and 2,6-dimethyl-(2*Z*,5*E*),7-octatrienyl acetate ((*E*,*Z*)-8-ocimenyl acetate) 9 seem to be very important flavor components for the flavor of peppermint oils. The trivial name we have given to the isomeric mixture is *piperitanate*. (*E*,*Z*)-8-Ocimenyl acetate 9 is the more relevant isomer of the two. It smells and tastes green, herbaceous, earthy, fruity and has distinct notes of pineapple and pine needle. It has a lower odor threshold (0.4 ppb/air) and was quantified in the Willamette oil with approximately 30 ppm. (*Z*,*Z*)-8-Ocimenyl acetate 7 smells and tastes herbaceous, galbanum-like, green, pineapple-like and fruity. It has an odor threshold of 5 ppb/air and was quantified in the Willamette oil with approximately 20 ppm. Both isomers remind us somewhat of another important flavor compound in piperita oils namely, 1,(3*E*,5*Z*)-undecatriene, the character impact compound of galbanum oil which smells typically green and earthy. The discussed isomeric ocimenyl acetates represent and support sensorically the "green, fruity, herbaceous" aspects of a peppermint oil. They also seem to boost the hay and sweet notes in flavor compositions.

It is very interesting to note that in the recent article by Benn on the important odorants in peppermint and cornmint oils (*7*) in Table 2 under #57 an unknown compound (retention indices 998/1215) with a high flavor dilution factor of 7 was described as "juniper, pineapple, green". Bearing in mind that the retention index system described in this paper used ethyl esters as reference standards instead of hydrocarbons used in our work it seems quite obvious to us that this unknown component is in fact 2,6-dimethyl-(2*Z*,5*E*),7-octatrienyl acetate ((*E*,*Z*)-8-ocimenyl acetate) 9. The analytical data as well as the sensory description fit quite well with our findings. Another unknown compound in Table 2 under #20 with a flavor dilution factor of 8 in this study by Benn that was described as "juniper, fruity, green, pineapple" must be 1,(3*E*,5*Z*)-undecatriene, according to our data.

Bicyclic Monoterpene Furans and Lactones

Another sensorially very important class of compounds in peppermint oil comprises the bicyclic monoterpene furans and lactones. The discussed compounds are listed in Table 3 along with some analytical parameters and their sensory characteristics. Very well-known are the so-called menthofuran (6R) 14, mintlactone (6R, 7aR) 16, isomintlactone (6R, 7aS) 17, and dehydromintlactone (6R) 18. All four bicyclic monoterpenes impart "coconut, sweet, hay-like and coumarin-like" aspects to a peppermint oil. Three stereoisomers of perhydro-3,6-dimethylbenzo-[b]-furan-2-one 20 were found and published for the first time recently by Näf et al. (*8*) in Italo-Mitcham peppermint oil. They also reported finding 3,6-dimethylbenzo-[b]-furan 15 and 3,6-dimethylbenzo-[b]-furan-2(3H)-one 19 in the same oil. A particularly interesting compound is 3,6-dimethylbenzo-[b]-furan-2(3H)-one 19. We had identified it in the Willamette oil, synthesized and tested for its flavor properties already when the above mentioned paper was published. The flavor properties of

Table II: Analytical Parameters and Sensory Characteristics of Some Acyclic Monoterpenes in Peppermint Oil (*Mentha piperita L.*)

Flavor Component	Retention Index DB-1/DB-Wax	Identified in Peppermint Oil	Sensory Impression	Odor Threshold/Taste Threshold
myrcene 1	985/1164	Willamette	terpeney, woody	112.5 ng/L air (ref. 41 ng/L air)
(*Z*)-β-ocimene 2	1027/1234	Willamette	terpeney	10 ng/L air
(*E*)-β-ocimene 3	1038/1252	Willamette	terpeney, mango-like	18.7 ng/L air
2-methyl-6-methylene-(2*Z*),7-octadienol	1206/1881	Willamette	floral, lily-of-the-valley	
2-methyl-6-methylene-(2*Z*),7-octadienyl acetate (*Z*-8-myrcenyl	1322/1763	Willamette	herbaceous, mango-like, terpeney	
2,6-dimethyl-(2*Z*,5*Z*),7-octatrienol (*Z*,*Z*-8-ocimenol) 6	1250/1960	Willamette	woody, musty, resin-like	
2,6-dimethyl-(2*Z*,5*Z*),7-octatrienyl acetate (*Z*,*Z*-8-ocimenyl acetate) 7	1363/1829	Willamette	herbaceous, galbanum-like, green, pineapple, fruity	5 ng/L air 1–2 μg/L 5% aqueous sugar solution
2,6-dimethyl-(2*Z*,5*E*),7-octatrienol (*E*,*Z*-8-ocimenol) 8	1264/1988	Willamette		
2,6-dimethyl-(2*Z*,5*E*),7-octatrienyl acetate (*E*,*Z*-8-ocimenyl acetate) 9	1375/1855	Willamette	green, herbaceous, earthy, fruity, pineapple, pine needle	0.4 ng/L air 7.8–125 μg/L 5% aqueous sugar solution

Table III: Analytical Parameters and Sensory Characteristics of Some Bicyclic Monoterpenes in Peppermint Oil (*Mentha piperita L.*)

Flavor Component	Retention Indices DB-1/DB-WAX	Identified in Peppermint Oil	Sensory Impression	Odor Threshold/Taste Threshold
menthofuran 14	1150/1485	Willamette	sweet, minty, lactoney, hay-like	0.4 ng/L air
3,6-dimethylbenzo-[b]-furan 15	1199/1689	Willamette		
mintlactone 16	1455/2310	Willamette	coconut, sweet, coumarin-like	2.8 ng/L air 1.9 – 7.5 µg/L 5% aqueous sugar solution
isomintlactone 17	1476/2338	Willamette	coconut, coumarin-like	1.25 ng/L air (ref. 3.5 ng/L air)
dehydromintlactone 18	1448/2255	Willamette	coconut, coumarin-like, hay-like	3.1 ng/L air 0.5 – 3.8 µg/L 5% aqueous sugar solution
3,6-dimethylbenzo-[b]-furan-2(3H)-one (furamintone) 19	1196/1689	Willamette	coconut, coumarin-like, sweet, lactoney	0.8 ng/L air 0.9 – 3.8 µg/L 5% aqueous sugar solution
perhydro-3,6-dimethyl-benzo-[b]-furan-2-one 20	1382/2134	Willamette		

furamintone, the trivial name we have given to this flavor compound, seem quite similar to the other bicyclic lactones but it has a lower odor threshold and is a much stronger odoriferous compound. The racemic mixture was described as being "coconut-like, coumarin-like and sweet". Its odor threshold was determined to 0.8 ng/L air. Its taste threshold and taste recognition threshold were 0.9 µg/L sugar solution and 3.8 µg/L sugar solution, respectively. We also determined the odor thresholds of the other bicyclic monoterpenes. Menthofuran has an odor threshold of 0.4 ng/L air, mintlactone of 2.8 ng/L air. The taste threshold and taste recognition threshold of mintlactone were determined as 1.9 µg/L and 7.5 µg/L, respectively. Isomintlactone has an odor threshold in air of 1.25 ng/L. Guth et al. just recently reported it as 3.5 ng/L air *(23,24)* which is in very good agreement with our finding.

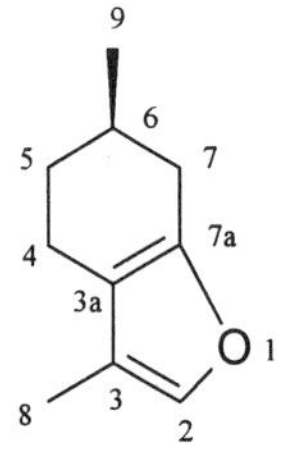

14 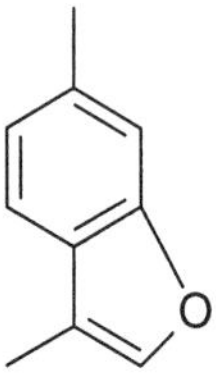 15 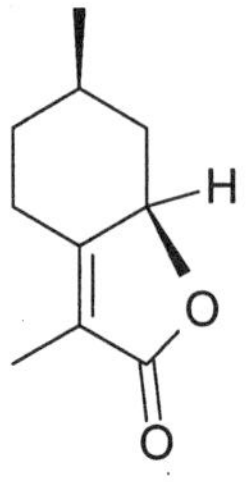 16 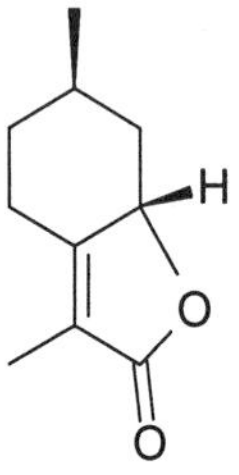 17

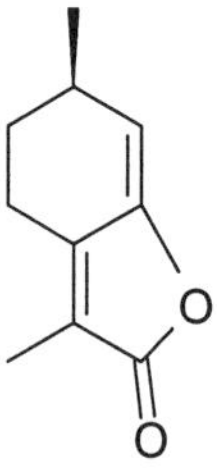

18 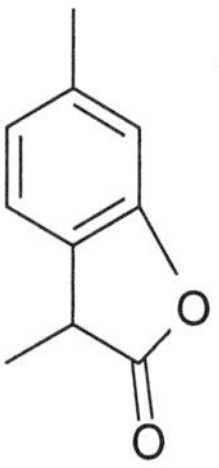 19 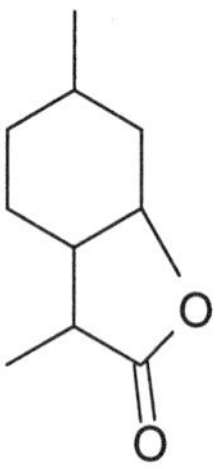 20

Finally, the thresholds of dehydromintlactone were also determined. The odor threshold in air is 3.1 ng/L and the taste and taste recognition thresholds are 0.5 µg/L and 3.8 µg/L sugar solution.

Some of the described bicyclic furans and lactones have one or several asymmetric carbon atoms. Consequently, several stereoisomers are possible. It has been shown (*3,6,25*) that the absolute configuration at C-6 of the concerned bicyclic compounds in peppermint is always related to the absolute configuration of C-1 of menthol (R-configurated). Consequently, (+)-(6R)-menthofuran 14, (-)-(6R,7aR)-mintlactone 16, and (+)-(6R)-dehydromintlactone 18 were identified enantiomerically pure in peppermint oil.

3,6-Dimethylbenzo-[b]-furan-2(3H)-one (furamintone) 19 has one asymmetric carbon atom (C-3) consisting consequently of the two stereoisomers (3S,6)-dimethylbenzo-[b]-furan-3(2H)-one 19a and (3R,6)-dimethylbenzo-[b]-furan-3(2H)-one 19b. By performing the gas chromatographic separation on a chiral phase (30% DMTBS-β-cyclodextrin, 70% PS086, length 25 m, ∅ 0.25 mm, film 0.25 μm, temperature program 80° – 150°C at a rate of 2°C/min, MEGA, Italy) with the synthesized racemic compound we found out that the separation of the two enantiomers was straightforward. The two enantiomers were baseline-separated. GC-olfactometry of the racemate using the chiral phase lead to the result that both enantiomers had the same sensory properties and were described as "coconut-like, coumarin-like and sweet." Their odor thresholds were also the same and determined to 0.8 ng/L air. The elution order though remains unknown since the absolute configuration of the two stereoisomers has not been determined yet. The enantiomeric ratio of furamintone 19 in peppermint oil should be 1:1 since its biogenetic origin is very likely related to dehydromintlactone 18.

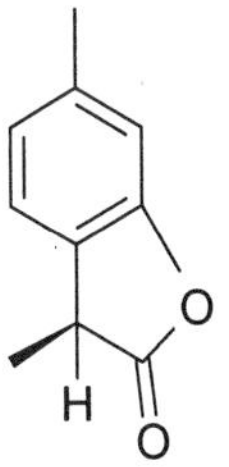

19a 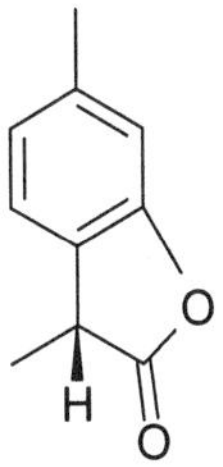 19b

Bicyclic monoterpene lactones are in general a very interesting and important class of flavor compounds. While the above mentioned mintlactone, isomintlactone, dehydromintlactone and furamintone are important for the flavor of peppermint oil especially one other structure has recently been published in the literature being a highly potent flavor compound in wine, orange and grapefruit.

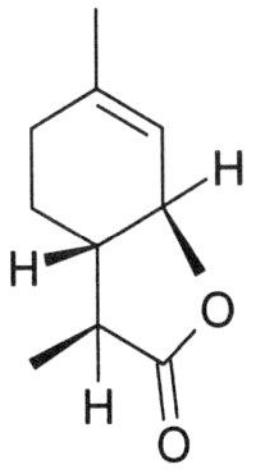

20 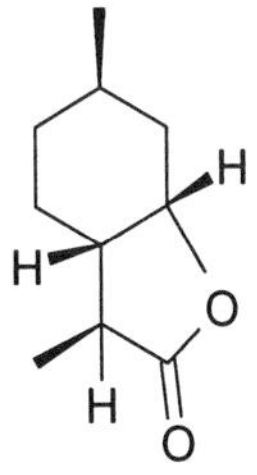 21

3,6-Dimethyl-3a,4,5,7a-tetrahydrobenzo-[b]-furan-2(3H)-one 20, the so-called wine lactone, was first identified by Guth et al. in wine (*26*). He synthesized and separated all the stereoisomers and identified the one in wine that has a very low odor threshold of 0.00002 ng/L air. Very recently this wine lactone was also found in orange and grapefruit flavor (*27-29*).

Another interesting finding was reported very recently. In systematic studies about the correlation of chemical structures with odor properties and odor thresholds Guth et al. (*23,24*) reported that the 3S,3aS,6R,7aS stereoisomer of perhydro-3,6-dimethylbenzo-[b]-furan-2-one 21 has the lowest odor threshold found so far for flavor components. It was determined as 0.000001 ng/L air. Interestingly, this stereoisomer has never described being found in food or essential oils so far.

In addition to the above listed thresholds of the acyclic monoterpenes and bicyclic monoterpenes, we determined the odor thresholds of a few more flavor chemicals that are important in peppermint oil (Table 4) and used them as references for our method (see Experimental Procedures). The odor threshold of 1,8-cineole (eucalyptol) was determined as 3.1 ng/L air. It correlated well with the threshold published in the literature (*30*). Very recently it was published as an important flavor volatile in black pepper with an odor threshold determined in starch of 84 μg/kg (*31*). Sensorically, it can be described as "fresh, eucalyptus-like, cool". The threshold of racemic linalool was determined as 2.0 ng/L air. It also correlated quite well with the published value in the literature (*30,32*). It can be described with „floral, fruity, citrus-like" notes. A very interesting result was obtained with (E)-β-damascenone. We determined the odor threshold of this important flavor compound with 0.125 ng/L air. Compared with the value of 0.003 ng/L air published in the literature (*30,32*) this is a difference of two orders of magnitude. Therefore we repeated the experiment several times and we also checked carefully the chemical itself. But the result remained the same. Therefore we have to report an odor threshold for (E)-β-damascenone that is not in agreement with the one published in literature. Finally, we also determined the odor threshold of 1,(3*E*,5*Z*)-undecatriene. It was determined with 0.125 ng/L air. To our knowledge the threshold of this important flavor compound has not been reported yet. It is the typical impact compound of galbanum roots and has "herbaceous, green" notes.

Acknowledgments

The authors would like to thank the entire groups of organic synthesis, chromatography, spectroscopy, flavor research, and flavor development for their valuable and skillful work. A special gratitude goes to our flavorist Volkmar Koppe and to Stephan Trautzsch.

Table IV: Analytical Parameters and Sensory Characteristics of Some Additional Flavor Components in Peppermint Oil (*Mentha piperita L.*)

Flavor Component	Retention Indices DB-1/DB-WAX	Identified in Peppermint Oil	Sensory Impression	Odor Threshold/Taste Threshold
(E)-β-damascenone	1362/1826	Willamette	flowery, tea-like, dry plum, herbaceous	0.125 ng/L air (ref. 0.002 – 0.004 ng/L air)
1,8-cineole (eucalyptol)	1025/1225	Willamette	fresh, cool, eucalyptus	3.1 ng/L air (ref. 3 ng/l air)
1,(3*E*,5*Z*)-undecatriene	1166/1393	Willamette	herbaceous, galbanum-like	0.125 ng/L air
linalool (racemic)	1087/1545	Willamette	floral, fruity, citrus-like	2 ng/L air (ref. 0.4 – 0.8 ng/L air)

References

1. Lawrence, B.M.; Shu, C-K. *Perfumer & Flavorist* **1989**, *14*, 21.
2. Lawrence, B.M. *Perfumer & Flavorist* **1993**, *18*, 59.
3. Faber, B.; Dietrich, A.; Mosandl, A. *J. Chromatogr.* **1994**, *666*, 161.
4. Faber, B.; Krause, B.; Dietrich, A.; Mosandl, A. *J. Essent. Oil Res.* **1995**, *7*, 123.
5. Spencer, J.S.; Dowd, E.; Faas, W. *Perfumer & Flavorist* **1997**, *22*, 37.
6. König, W.A.; Fricke, C.; Saritas, Y.; Momeni, B.; Hohenfeld, G. *J. High Resol. Chromatogr.* **1997**, *20*, 55.
7. Benn, S. *Perfumer & Flavorist* **1998**, *23*, 5.
8. Näf, R.; Velluz, A. *Flav. Fragr. J.* **1998**, *13*, 203.
9. Güntert, M.; Bruening, J.; Emberger, R.; Koepsel, M.; Kuhn, W.; Thielmann, T.; Werkhoff, P. *J. Agric. Food Chem.* **1990**, *38*, 2027.
10. Güntert, M.; Brüning, J.; Emberger, R.; Hopp, R.; Köpsel, M.; Surburg, H.; Werkhoff, P. In *Flavor Precursors – Thermal and Enzymatic Conversions*; Teranishi, R.; Takeoka, G.R.; Güntert, M., Eds.; ACS Symposium Series 490, American Chemical Society: Washington DC, 1992, pp 140–163.
11. Werkhoff, P.; Güntert, M.; Krammer, G.; Sommer, H.; Kaulen, J. *J. Agric. Food Chem.* **1998**, *46*, 1076.
12. Ullrich, F.; Grosch, W. *Z. Lebensm. Unters. Forsch.* **1987**, *184*, 277.
13. Ullrich, F.; Grosch, W. *JAOCS* **1988**, *65*, 1313.
14. Granger, R.; Passet, J.; Girard, J.P. *Phytochemistry* **1972**, *11*, 2301.
15. Passet, J. *Parfums, Cosmet., Aromes* **1979**, *28*, 39.
16. Delpit, B.; Lamy, J.; Rolland, F.; Chalchat, J.C.; Garry, R.Ph. *Riv. Ital. EPPOS* **1996**, *7*, 409.
17. Adegoke, G.O.; Jagan Mohan Rao, J.; Shankaracharya, N.B. *Flavour Fragr. J.* **1998**, *13*, 349.
18. Güntert, M.; Krammer, G.; Sommer, H.; Werkhoff, P. In *Flavor Analysis – Developments in Isolation and Characterization*; Mussinan, C.J.; Morello, M.J., Eds.; ACS Symposium Series 705, American Chemical Society: Washington DC, 1998, pp 38–60.
19. Toulemonde, B.; Noleau, I. In *Flavors and Fragrances – A World Perspective*; Lawrence, B.M.; Mookherjee, B.D.; Willis, B.J., Eds.; Proceedings of the 10th International Congress of Essential Oils, Fragrances and Flavors, Washington, DC, U.S.A., 16 – 20 November 1986, Elsevier Science Publishers B.V.: Amsterdam, 1988, pp 641-657.
20. Strauss, C.R.; Wilson, B.; Williams, P.J. *J. Agric. Food Chem.* **1988**, *36*, 569.
21. Mandery, H. Diploma Thesis, University of Karlsruhe, Germany, 1985.
22. Bock, B.; Benda, I.; Schreier, P. In *Biogeneration of Aromas*; Parliment, T.H.; Croteau, R., Eds.; ACS Symposium Series 317, American Chemical Society: Washington DC, 1986, pp 243-253.
23. Guth, H.; Buhr, K.; Fritzler, R. Lecture No. 21, Weurman Symposium, June 22–25, 1999, Freising, Germany.

24. Guth, H.; Buhr, K.; Fritzler, R. Lecture No. 15, Division of Agricultural & Food Chemistry, American Chemical Society Meeting, August 22-26, 1999, New Orleans.
25. Werkhoff, P; Brennecke, S; Bretschneider, W.; Güntert, M.; Hopp, R.; Surburg, H. *Z. Lebensm. Unters. Forsch.* **1993**, *196*, 307.
26. Guth, H. *Helv. Chim. Acta* **1996**, *79*, 1559.
27. Hinterholzer, A.; Schieberle, P. *Flavour Fragr. J.* **1998**, *13*, 49.
28. Schieberle, P.; Büttner, A. Lecture No. 4, Weurman Symposium, June 22–25, 1999, Freising, Germany.
29. Büttner, A.; Schieberle, P. Poster No. 8, Weurman Symposium, June 22–25, 1999, Freising, Germany.
30. Rychlik, M.; Schieberle, P.; Grosch, W. Compilation of odor thresholds, odor qualities and retention indices of key food odorants. Deutsche Forschungsanstalt für Lebensmittelchemie der Technischen Universität München, 1988, p.10 and p.28.
31. Jagella, T.; Grosch, W. *Eur. Food Res. Technol.* **1999**, *209*, 22.
32. Blank, I.; Fischer, K.-H.; Grosch, W. *Z. Lebensm. Unters. Forsch.* **1989**, *189*, 426.

Chapter 11

Chemical and Sensory Properties of Thiolactones

K.-H. Engel, A. Schellenberg, and H.-G. Schmarr

Technische Universität München, Lehrstuhl für Allgemeine Lebensmitteltechnologie, Am Forum 2, D–85350 Freising, Germany

Model experiments were performed to determine the recoveries of γ-/δ-thiolactones by isolation from aqueous solutions and the stabilities of these volatiles upon heat treatment at different pH-values. Using simultaneous distillation-extraction, thiolactones exhibited higher recoveries than the corresponding lactones. Upon refluxing at neutral or alkaline conditions and subsequent liquid-liquid extraction, thiolactones turned out to be unstable. Using octakis-(2,3-di-*O*-acetyl-6-*O*-*tert*-butyldimethylsilyl)-γ-cyclodextrin as stationary phase for capillary gas chromatography, a baseline separation for γ- as well as δ-thiolactones could be achieved. Enantioselective enzyme-catalyzed hydrolysis of δ-thiooctalactone using *Porcine Pancreas* lipase proceeded enantioselectively, resulting in (R)-δ-thiooctalactone and (S)-5-mercaptooctanoic acid of high optical purity. GC-olfactometry was applied to describe odor properties of thiolactone enantiomers and derivatives thereof.

Introduction

Flavor and aroma compounds belong to a diverse spectrum of chemical substances *(1)*. Sulfur-containing volatiles are among the most prominent and important representatives *(2)*. They are often characterized by a combination of low threshold and pronounced odor properties *(3)*. Accordingly, many of the so-called "character impact compounds" of foods are sulfur-containing substances. They decisively influence thermally produced flavors *(4)* and they also play outstanding roles in the patterns of biosynthesized aroma constituents of several fruits, such as passion fruit *(5)*, blackcurrant *(6)* and orange *(7)*.

In many cases the substitution of the hydroxyl group by a mercapto group results in drastic changes of the sensory properties of a compound *(3)*. An impressive example for this effect is the difference in odor quality and threshold between α-terpineol and p-menthene-8-thiol *(8)*.

In recent studies γ- and δ-thiolactones have been synthesized in order to investigate the effect of a replacement of oxygen in the ring of aliphatic lactones by sulfur on the sensory properties. The thiolactones turned out to have attractive sensory properties; especially δ-thiolactones exhibit a combination of the warm lactone character and spicy, tropical fruit notes *(9, 10)*.

In this contribution additional data on the analytical and sensorial characterization of γ- and δ-thiolactones are presented. The following aspects will be covered: (i) chemical stability and recovery, (ii) capillary gas chromatographic enantiodifferentiation, (iii) enzyme-catalyzed kinetic resolution of the enantiomers, and (iv) odor properties of the thiolactones and derivatives thereof.

Experimental

Stability and Recovery

Two model mixtures were prepared containing 20 mg each of γ- and δ-lactones and the corresponding thiolactones, respectively. Undecan-2-one (20 mg) was used as internal standard. The substances were dissolved in 5 ml *n*-hexane and filled up to 20 ml with ethanol. From this stock solution a dilution in ethanol (200 ng/μl) was prepared and used for the experiments. 1 ml of the diluted model mixture and 9 ml buffer (citrate/HCl for pH 3.5, phosphate for pH 7.0 and glycine/NaOH for pH 9.0) were subjected to the following isolation procedures: (i) micro simultaneous distillation-extraction *(11)* for 20 min using *n*-pentane/diethyl ether (1:1, v:v) as solvent, (ii) extraction with diethyl ether after stirring for 20 min at room temperature, and (iii) extraction with diethyl ether after refluxing for 20 min. The extracts were concentrated to 1 ml using a Vigreux column and analyzed by capillary gas chromatography.

Capillary Gas Chromatographic Enantioseparation

Separation of the enantiomers was achieved on a fused silica column (30 m x 0.25 mm i.d.) coated in the laboratory with 50% octakis-(2,3-di-*O*-acetyl-6-*O*-*tert*-butyldimethylsilyl)-γ-cyclodextrin *(12)* in OV-1701-vi (Supelco, Germany) to provide a film thickness of 0.25 μm. The column was installed in a gas chromatograph (Carlo Erba model 6000, Thermoquest, Germany) equipped with a flame ionization detector (FID) and a split injector. Hydrogen was used as carrier gas at a constant inlet pressure of 100 kPa. The oven temperature was programmed from 100°C (2 min hold) to 125°C (10 min hold) with 2°/min and then to 205°C (10 min hold) with 3°/min. Data aquisition was with the Chromcard system (Thermoquest).

Gas Chromatography-Olfactometry (GC/O)

The odor qualities of the compounds were determined by GC/O injecting about 100 ng onto the above mentioned column. The column was installed in a modified GC model HP 5880 (Hewlett Packard, Germany) equipped with on-column injector and FID. Hydrogen was used as carrier gas at a constant inlet pressure of 100 kPa. The column effluent was split 1:1 via a press-fit Y-piece to a heated sniffing port and the FID. Temperature was programmed from 40°C (2 min hold) to 120°C (1 min hold) with 40°/min and then to 160°C (2 min hold) with 1°/min. Flavor dilution factors (FD) of the compounds were determined by aroma extract dilution analysis (AEDA) *(13, 14)*. After stepwise dilutions (1:1) with diethyl ether, aliquots were analyzed by GC/O. Odor thresholds were determined as described *(13)* using (E)-2-decenal as internal standard *(15)*.

Enzymatic Hydrolysis of δ-Thiolactones

Kinetic resolution of the enantiomers of δ-thiolactones was performed via hydrolysis in a buffered aqueous medium catalyzed by *Porcine Pancreas* lipase (Sigma, Germany) *(16)*. Experimental conditions were as follows: 5 ml phosphate buffer (pH 8), 16 mg lipase, 2.5 μl nonan-2-one (internal standard) and 0.12 mmol thiolactone were reacted at 30°C. Aliquots were taken at different time intervals, extracted with diethyl ether and analyzed by gas chromatography.

Results and Discussion

Stability and Recovery

Owing to the attractive sensory properties of γ- and δ-thiolactones, it seems worthwhile to put efforts into a search for the natural occurrence of these volatiles. A prerequisite for such investigations is the application of appropriate isolation techniques and knowledge of the chemical stabilities of the target compounds. Tables I-III summarize results from model experiments performed to determine the recoveries of γ-/δ-thiolactones by isolation from aqueous solutions and the stabilities of these volatiles upon heat treatment at different pH-values. Using micro simultaneous distillation-extraction (SDE) as isolation procedure, thiolactones generally exhibit higher recoveries than the corresponding lactones (Table I). The replacement of oxygen by sulfur apparently reduces the hydrophilic character of the compounds resulting in higher enrichment in the vapor phase and accordingly increased yield in the course of the extraction step.

The effect of the differences in polarities between oxygen and sulfur is especially reflected in the behavior of γ- and δ-thiohexalactone. In contrast to the higher homologs, the lipophilic character of the C_6-compounds is not mainly determined by the alkyl chain and thus more influenced by the heteroatom in the ring. In accordance with data reported for oxygen containing lactones *(17)*, the recoveries of higher homologs increase with increasing chain lengths, due to increasing lipophilicity.

SDE at pH 7 and 9, respectively, reveals the low stability of lactones as well as thiolactones at higher pH-values. Owing to their reduced hydrophilic character, a faster release of the thiolactones from the aqueous phase occurs and due to the shorter exposure to unfavorable pH-conditions, their recoveries are higher than those of the oxygen containing homologs.

Liquid-liquid extraction at pH 3 and 7, respectively, did not show significant differences in recoveries for lactones and thiolactones (Table II). Only at pH 9 the differences in stability between the 5- and the 6-membered ring homologs became obvious for the thiolactones.

Table I. Recoveries of Lactones and Thiolactones from Aqueous Solutions by Micro Simultaneous Distillation-Extraction

	pH 3.5 %[c]		*pH 7.0* %[c]		*pH 9.0* %[c]	
	O[a]	*S*[b]	*O*	*S*	*O*	*S*
γ-lactones						
C_6	88 ± 4	100 ± 1	55 ± 12	102 ± 3	44 ± 5	85 ± 11
C_8	99 ± 1	96 ± 1	98 ± 9	101 ± 8	79 ± 3	87 ± 6
C_{10}	98 ± 1	100 ± 1	105 ± 14	108 ± 13	84 ± 10	94 ± 5
δ-lactones						
C_6	8 ± 2	90 ± 1	3 ± 2	61 ± 3	1 ± 1	12 ± 4
C_8	56 ± 6	96 ± 2	27 ± 6	82 ± 1	10 ± 6	28 ± 5
C_{10}	93 ± 1	100 ± 1	66 ± 4	92 ± 1	34 ± 6	52 ± 3

[a] oxygen containing lactones, [b] thiolactones, [c] mean values ± standard deviations of three experiments.

This effect was even more pronounced upon refluxing at neutral or alkaline conditions and subsequent liquid-liquid extraction (Table III). Recoveries after refluxing for 20 min are generally lowered compared to direct extraction. Thiolactones turned out to be more susceptible to heat treatment at higher pH. The lower stability of the 6-membered rings is especially obvious for the sulfur-containing homologs.

Table II. Recoveries of Lactones and Thiolactones from Aqueous Solutions by Liquid-Liquid Extraction

	pH 3.5 %[c]		*pH 7.0* %[c]		*pH 9.0* %[c]	
	O[a]	*S*[b]	*O*	*S*	*O*	*S*
γ-lactones						
C_6	108 ± 1	112 ± 1	110 ± 2	114 ± 2	113 ± 5	112 ± 5
C_8	111 ± 1	107 ± 1	114 ± 2	109 ± 2	117 ± 4	108 ± 4
C_{10}	113 ± 1	113 ± 3	112 ± 2	114 ± 2	115 ± 5	114 ± 4
δ-lactones						
C_6	75 ± 5	106 ± 1	87 ± 4	99 ± 2	77 ± 1	23 ± 2
C_8	83 ± 8	139 ± 7	99 ± 2	103 ± 2	93 ± 1	40 ± 3
C_{10}	89 ± 9	118 ± 1	105 ± 2	112 ± 1	100 ± 2	86 ± 4

[a] oxygen containing lactones, [b] thiolactones, [c] mean values ± standard deviations of three experiments.

Table III. Recoveries of Lactones and Thiolactones from Aqueous Solutions after Refluxing for 20 min and Subsequent Liquid-Liquid Extraction

	pH 3.5 %[c]		*pH 7.0* %[c]		*pH 9.0* %[c]	
	O[a]	*S*[b]	*O*	*S*	*O*	*S*
γ-lactones						
C_6	85 ± 9	68 ± 5	78 ± 6	74 ± 5	60 ± 0	42 ± 4
C_8	84 ± 5	59 ± 3	79 ± 6	69 ± 4	63 ± 1	42 ± 4
C_{10}	78 ± 9	50 ± 3	80 ± 6	60 ± 5	63 ± 0	41 ± 4
δ-lactones						
C_6	37 ± 5	72 ± 13	32 ± 4	29 ± 3	6 ± 1	0.3 ± 0.1
C_8	50 ± 6	92 ± 15	54 ± 4	39 ± 4	20 ± 3	3 ± 0
C_{10}	58 ± 7	74 ± 12	69 ± 6	39 ± 4	27 ± 3	4 ± 1

[a] oxygen containing lactones, [b] thiolactones, [c] mean values ± standard deviations of three experiments.

Enantioseparation

A first separation of the enantiomers of γ- and δ-thiolactones had been reported on heptakis-(2,3-di-*O*-methyl-6-*O*-*tert*-butyldimethylsilyl)-β-cyclodextrin as chiral stationary phase *(9)*. In order to improve the low resolution of the δ-thiolactone enantiomers, other modified chiral cyclodextrin derivatives were tested. As shown in Figure 1, the use of octakis-(2,3-di-*O*-acetyl-6-*O*-*tert*-butyldimethylsilyl)-γ-cyclodextrin as stationary phase results in baseline separation of γ- as well as δ-thiolactone enantiomers.

The assignment of the elution order of the enantiomers was based on investigations of the 4- and 5-mercaptoalcohols obtained after reductive cleavage of the thiolactones with $LiAlH_4$. A sequence involving resolution of the mercaptoalcohol enantiomers via HPLC and ^{1}H-NMR spectroscopy of diastereomeric (R)-2-phenylpropionic esters allowed the determination of the absolute configurations *(18)*.

Enzyme-Catalyzed Kinetic Resolution of δ-Thiolactone Enantiomers

Enzyme-catalyzed reactions are useful approaches for kinetic resolutions of the enantiomers of chiral substances *(19-21)*. For lactones, lipase-catalyzed transesterification in organic solvents *(22)* as well as enzyme-mediated hydrolyses in aqueous medium *(23, 24)* have been described as useful strategies to obtain optically enriched compounds. In the course of a screening of commercially available enzyme preparations for their abilities to accept thiolactones as substrates and to catalyze enantioselective reactions, *Porcine Pancreas* lipase turned out to be a suitable biocatalyst. Hydrolyses of racemic δ-thiolactones in buffered aqueous solution proceeded with a high preference for the (S)-enantiomer yielding optically enriched (S)-5-mercaptoalkanoic acid as product and (R)-thiolactone as remaining substrate. As summarized in Figure 2, reaction rate and enantioselectivity are influenced by the chain length of the starting δ-thiolactone. The most pronounced enantiodiscrimination was observed for the C_8 homolog. A practical example of the product spectrum obtainable by this approach is depicted in Figure 3. The mercaptoacid and thiolactone can be separated using sodium *p*-hydroxymercuribenzoate as reagent *(25)*. Subsequent recyclization of the (S)-5-mercaptoacid to the (S)-δ-thiolactone offers the possibility to obtain both thiolactone enantiomers in optically enriched form starting from the racemic material *(26)*.

GC-Olfactometry

δ-Thiolactone enantiomers obtained by the described enzyme-catalyzed kinetic resolution were used for sensory evaluations (Figure 4). The (S)-

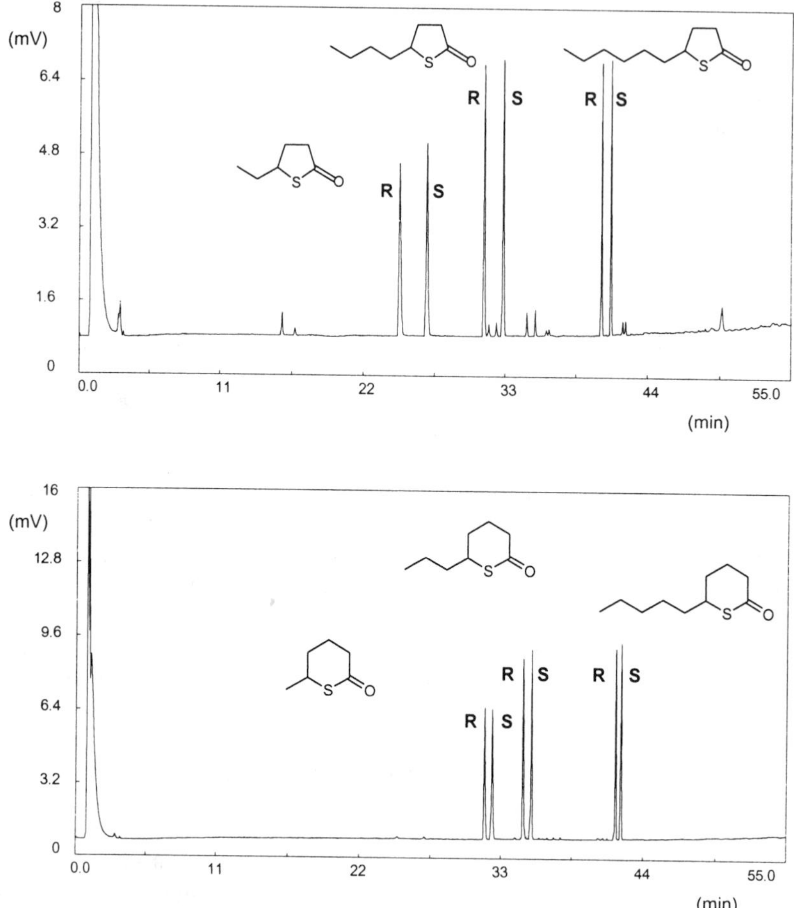

Figure 1. Capillary gas chromatographic enantiodifferentiation of γ- and δ-thiolactones (conditions see Experimental).

enantiomers of δ-thiooctalactone and δ-thiodecalactone exhibited slightly lower odor thresholds than the (R)-antipodes. For both lactones, significant differences in odor qualities were observed between the enantiomers. However, for the two examples studied, there was no consistent correlation between the tropical, fruity note and the configuration.

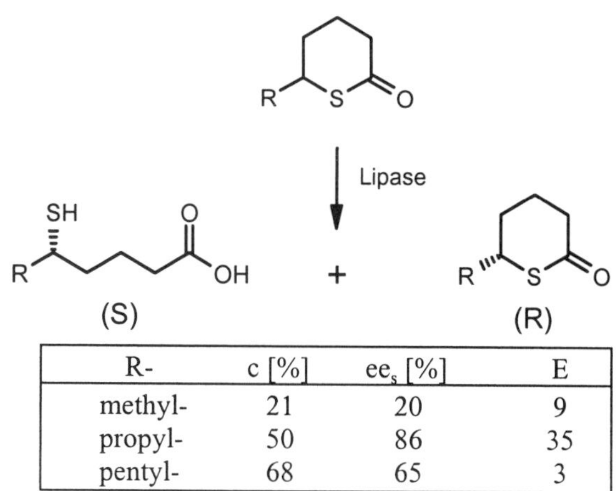

R-	c [%]	ee_s [%]	E
methyl-	21	20	9
propyl-	50	86	35
pentyl-	68	65	3

Figure 2. Stereochemical course of the kinetic resolution of δ-thiolactones via hydrolysis catalyzed by lipase from Porcine Pancreas. Determination of the extent of conversion (c), enantiomeric excess (ee_s) of the recovered substrate and enantiomeric ratio (E) according to Chen et al. (27).

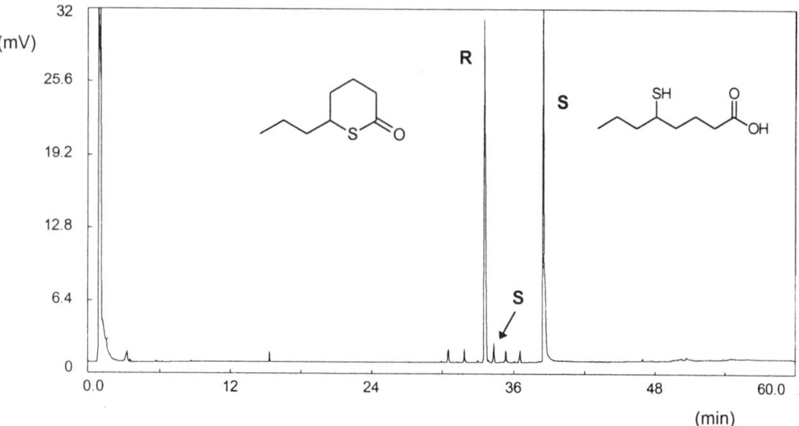

Figure 3. GC chromatogram of 5-mercaptooctanoic acid and δ-thiooctalactone obtained after enzymatic hydrolysis of racemic δ-thiooctalactone with Porcine Pancreas lipase (pH 8, 30°C, 1.5h); conditions as described in Experimental.

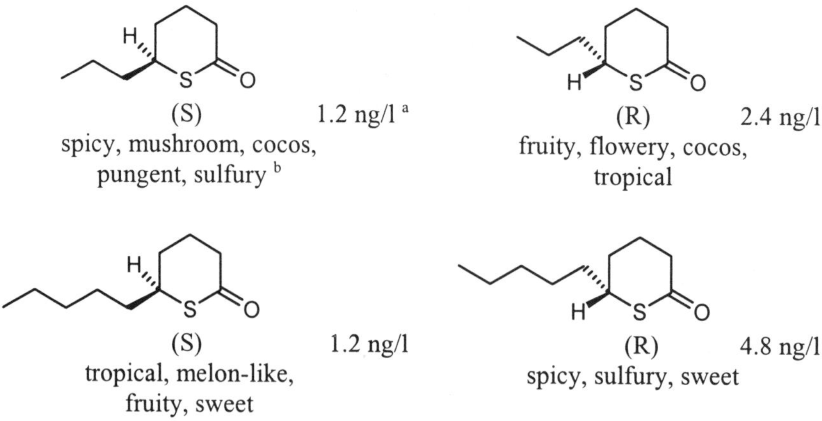

H S O

(S) 1.2 ng/l [a]
spicy, mushroom, cocos, pungent, sulfury [b]

H S O

(R) 2.4 ng/l
fruity, flowery, cocos, tropical

H S O

(S) 1.2 ng/l
tropical, melon-like, fruity, sweet

H S O

(R) 4.8 ng/l
spicy, sulfury, sweet

Figure 4. Sensory properties of δ-thiolactone enantiomers. [a] Odor thresholds in air. [b] GC/O description with 25 ng for each enantiomer at the sniffing port.

Sensory properties of mercaptoalcohols and mercaptoalkyl acetates, intermediates in the reaction sequence applied to determine the absolute configurations of the thiolactones *(18)*, were also evaluated. As example, Table IV lists odor descriptions obtained by GC/O for γ-thioheptalactone, 4-mercaptoheptanol and 4-mercaptoheptyl acetate. For the acetate significant differences in odor qualities between the enantiomers were noted. Systematic studies on the influence of the position of the mercapto group on the sensory properties of these compounds are part of ongoing research *(28)*.

Table IV. Comparison of Odor Qualities of γ-Thioheptalactone, 4-Mercapto-heptanol and 4-Mercaptoheptyl Acetate

	(S)	*(R)*
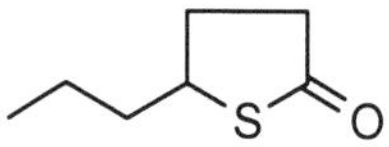 S O	cocos, sweet, spicy [a]	cocos, sweet, lacton, fruity
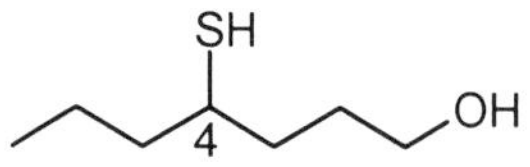 SH 4 OH	green, tropical, sulfury	green, tropical, burnt, sulfury
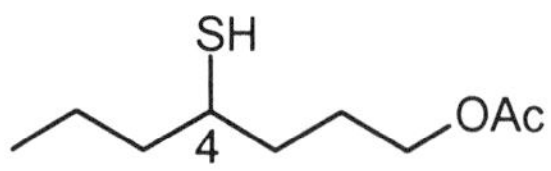 SH 4 OAc	mushroom, musty	fruity, tropical

[a] GC/O description with 25 ng for each enantiomer at the sniffing port.

References

1. *Volatile Compounds in Food: Qualitative and Quantitative Data*, 7th ed.; Nijssen, L. M., Ed.; TNO Nutrition and Food Research Institute: Zeist, Netherlands, 1996.
2. *Sulfur Compounds in Foods*; Mussinan, C. J.; Keelan, M. E., Eds.; ACS Symposium Series 564; American Chemical Society: Washington, DC, 1994.
3. Boelens, H. M. *Perfum. Flavor.* **1993**, *18*, 29-39.
4. Güntert, M.; Bertram, H.-J.; Hopp, R.; Silberzahn, W.; Sommer, H.; Werkhoff, P. In *Recent Developments in Flavor and Fragrance Chemistry*; Hopp, R.; Mori, K., Eds.; VCH: Weinheim, Germany, 1993; pp 215-240.
5. Engel, K.-H.; Tressl, R. *J. Agric. Food Chem.* **1991**, *39*, 2249-2252.
6. Rigaud, J.; Etievant, P.; Henry, R.; Latrasse, A. *Sciences des Aliments* **1986**, *6*, 213-220.
7. Büttner, A.; Schieberle, P. *J. Agric. Food Chem.* **1999**, *47*, 5189-5193.
8. Demole, E.; Enggist, P.; Ohloff, G. *Helv. Chim. Acta* **1982**, *65*, 1785-1794.
9. Roling, I.; Schmarr, H.-G.; Eisenreich, W.; Engel, K.-H. *J. Agric. Food Chem.* **1998**, *46*, 668-672.
10. Engel, K.-H.; Roling, I.; Schmarr, H.-G. In *Flavor Analysis: Developments in Isolation and Characterization*; Mussinan, C. J.; Morello, M. J., Eds.; ACS Symposium Series 705; American Chemical Society: Washington, DC, 1997; pp 141-151.
11. Godefroot, M.; Sandra, P.; Verzele, M. *J. Chromatogr.* **1981**, *203*, 325-335.
12. Schmarr, H.-G. Ph.D. thesis, J. W. Goethe University of Frankfurt, Frankfurt/Main, Germany, 1992.
13. Ullrich, F.; Grosch, W. *Z. Lebensm. Unters. Forsch.* **1987**, *184*, 277-282.
14. Schieberle, P.; Grosch, W. *Z. Lebensm. Unters. Forsch.* **1987**, *185*, 111-113.
15. Blank, I.; Fischer, K.-H.; Grosch, W. *Z. Lebensm. Unters. Forsch.* **1989**, *189*, 426-433.
16. Roling, I. Ph.D. thesis, Technical University of Munich, Munich, Germany, 1999.
17. Lehmann, D.; Dietrich, A.; Schmidt, S.; Dietrich, H.; Mosandl, A. *Z. Lebensm. Unters. Forsch.* **1993**, *196*, 207-213.
18. Schellenberg, A.; Schmarr, H.-G.; Eisenreich, W.; Engel, K.-H. In *Frontiers of Flavour Science*; Schieberle, P.; Engel, K.-H., Eds.; Deutsche Forschungsanstalt für Lebensmittelchemie: Garching, Germany, 2000; pp 121-124.
19. *Enzymatic reactions in organic media;* Koskinen, A. M. P.; Klibanov, A. M., Eds.; Blackie Academic & Professional: Glasgow, 1996.
20. Engel, K.-H. In *Flavor Precursors: Thermal and Enzymatic Conversions;* Teranishi, R.; Takeoka, G. R.; Güntert, M., Eds.; ACS Symposium Series 490; American Chemical Society: Washington, DC, 1992; pp 21-31.
21. Lutz, D.; Huffer, M.; Gerlach, D.; Schreier, P. In *Flavor Precursors: Thermal and Enzymatic Conversions;* Teranishi, R.; Takeoka, G. R.; Güntert, M., Eds.; ACS Symposium Series 490; American Chemical Society: Washington, DC, 1992; pp 32-45.
22. Huffer, M.; Schreier, P. *Tetrahedron Asymmetry* **1991**, *2*, 1157-1164.

23. Enzelberger, M. M.; Bornscheuer, U. T.; Gatfield, I.; Schmid, R. D. *J. Biotechnol.* **1997**, *56*, 129-133.
24. Blanco, L.; Guibé-Jampel, E.; Rousseau, G. *Tetrahedron Lett.* **1988**, *29*, 1915-1918.
25. Darriet, P.; Tominaga, T.; Lavignie, V.; Boidron, J.-N.; Dubourdieu, D. *Flav. Fragr. J.* **1995**, *10*, 385-392.
26. Schellenberg, A. Ph.D. thesis, Technical University of Munich, Munich, Germany, in prep.
27. Chen, C.-S.; Fujimoto, Y.; Girdaukas, G.; Sih, J. S. *J. Am. Chem. Soc.* **1982**, *104*, 7294-7299.
28. Schellenberg, A.; Schmarr, H.-G.; Eisenreich, W.; Engel, K.-H., unpublished.

Synthetic, Thermal Reaction, and Enzymatic Approaches to Flavor Components

Chapter 12

Synthetic Approaches to Chiral Flavor Components

Wilhelm Pickenhagen

Corporate Research, DRAGOCO Gerberding & Company AG, D–37603 Holzminden, Germany

Odor receptors are chiral, their interaction with chiral sensory compounds should be different with the two enantiomeric forms of those. Enantioselective methods are an important field of modern synthetic chemistry, allowing the selective synthesis of the optical isomers of flavor compounds. This paper discusses some general approaches to enantioselective syntheses and their application to some selected flavor impact compounds.

Introduction

It is established that the first event of the cascade that leads to an odor impression in vertebrates is the interaction of a stimulus, i.e. a volatile chemical entity, with a proteinaceous receptor (*1*). The exact nature of this interaction is still unknown. It can, however, be assumed that a certain fit, sterically and electronically, between parts of the receptor and the stimulus has to exist to allow the formation of a complex between the two. Proteins are chiral. It follows that this interaction will lead to two diastereomerically different complexes if the stimulus is a chiral compound, and these complexes that initiate the cascade could lead to different odor impressions.

Many sensorily active compounds are chiral (*2*). To evaluate the activity of their enantiomeric forms they have to be isolated. This can be done by separation or by selective syntheses.

Development of methods for asymmetric synthesis has become fashionable in the last fifteen years and some of them have matured to become industrially applicable.

Synthesis of Important Flavor Compounds

The following is a selection of these methods and their application to the synthesis of the enantiomeric forms of important flavor components (scheme 1).

Enantiomeric Agents That Are Lost After Reaction

n-Propyl-3-methyloxathiane **1** (Figure 1) has been identified as a flavor impact compound of the yellow passionfruit (*3*). For the enantioselective synthesis of both forms, trans-hex-2-enol is epoxidized enantioselectively (*4*) using the well-known Sharpless Reaction (*5*) (scheme 2).

Treatment of the epoxides with thiourea yields the optically active thiiranes of the opposite absolute configuration. This Walden Inversion at both chiral centers occurs as explained in scheme 3. Subsequent reactions (scheme 2) lead to the two enantiomeric forms of **1**. The organoleptic properties of these two compounds are quite different. **2**, having an odor threshold in water of 2 ppb, exhibits the typical sulfury notes of tropical fruits, whereas **3**, with a threshold of 4 ppb is slightly camphoracious, woody and has no tropical notes.

Enantiomeric Pure Starting Materials:

Menth-2-en-8-thiol **4** (Figure 2) has been found to occur in traces in grapefruit juice (*6*). It is described as one of the most powerful naturally occurring flavor compounds having an odor threshold in water of $t = 5 \times 10^{-4}$ ppb. Synthesis of both enantiomeric forms is straightforward using (-)- and (+)-limonene as starting material (scheme 4).

The two enantiomeric forms of **4** do not show significant differences in odor quality and strength.

Enantiomeric Agents That Can Be Recovered:

2-Methyl-3-mercaptopentanol **5** (Figure **3**) has recently been identified as a very powerful aroma component of fresh onions (*7*).

Selective synthesis of the four diastereo- and enantiomeric forms of **5** (*8*) uses Evans' benzyloxazolidinone procedure (*9*) for the induction of chirality (scheme 5, scheme 5a, scheme 5b).

Organoleptic analysis shows that the two threo (anti) forms **6** and **7** (Figure 4) having odor thresholds in water of 0.03 respectively 0.04 ppb are responsible for the powerful broth like, sweaty odor of this compound.

Separation Of Intermediate Diastereoisomers:

(-)-Menthol **8** (Figure 5) is a major constituent of mint oils and very much appreciated for its cooling effect, a property for which only the one enantiomeric form indicated in **8** is responsible. These cooling properties make (-)-menthol a very important flavor ingredient for which two enantioselective industrial syntheses have been developed.

The Haarmann & Reimer synthesis applies a selective crystallization of the benzoates for the separation of the two enantiomeric forms (scheme 6) (*10*).

Some Methods for Enantioselective Syntheses of Flavor Compounds

- ➡ use of enantiomeric agents that are lost after reaction
- ➡ enantiomeric pure starting material
- ➡ use of enantiomeric agents that can be recovered after reaction
- ➡ separation of intermediate diastereoisomers
- ➡ use of enantioselective catalysts
- ➡ use of enzymes

Scheme 1. Some Methods for Enantioselective Syntheses of Flavor Compounds

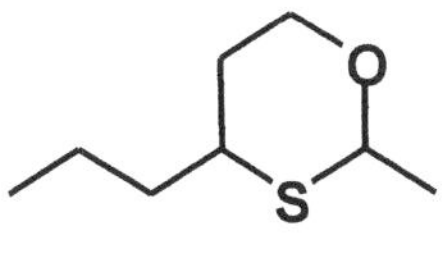

O S

Figure 1.

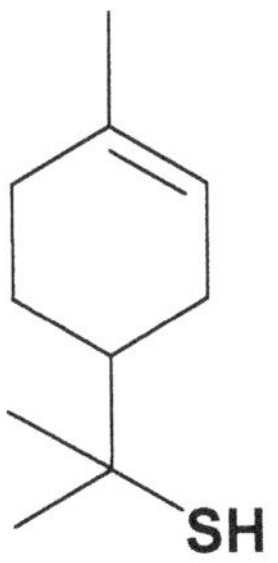

SH

Figure 2.

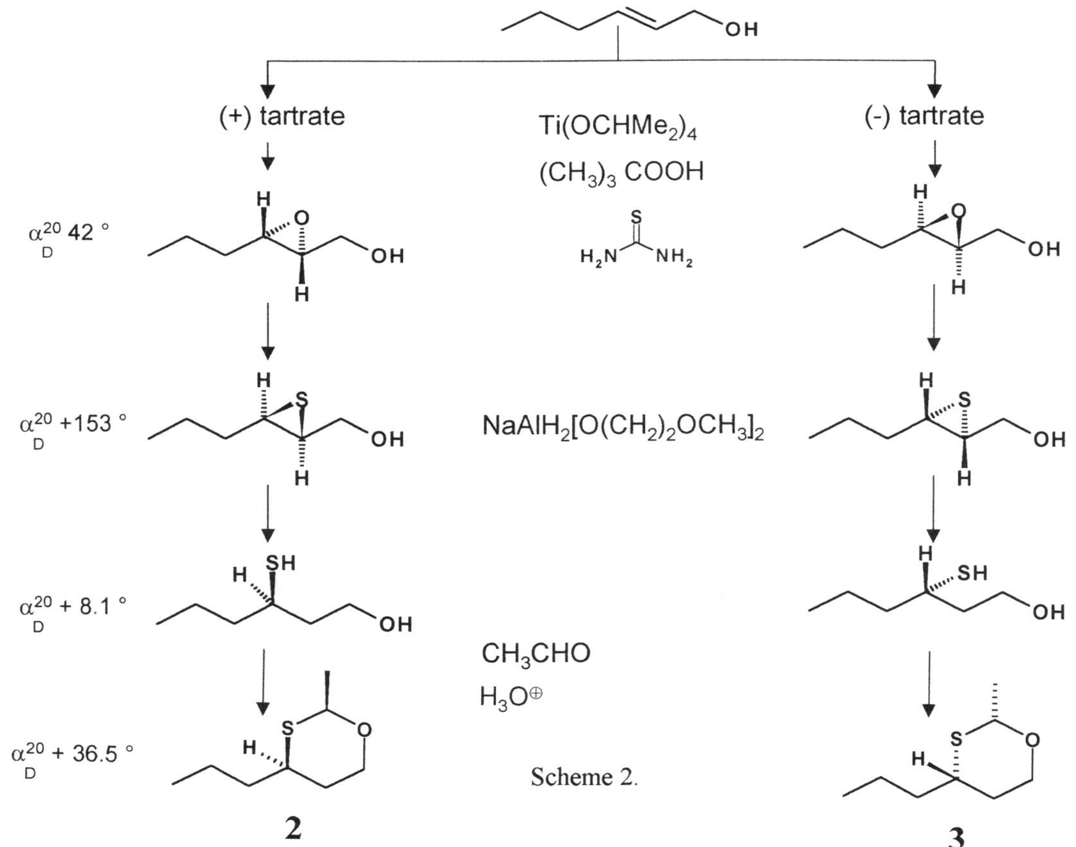

(+) tartrate
(-) tartrate
Ti(OCHMe2)4
(CH3)3 COOH
NaAlH2[O(CH2)2OCH3]2
CH3CHO
H3O⊕
α20D 42 °
α20D +153 °
α20D + 8.1 °
α20D + 36.5 °
2
3

Scheme 2.

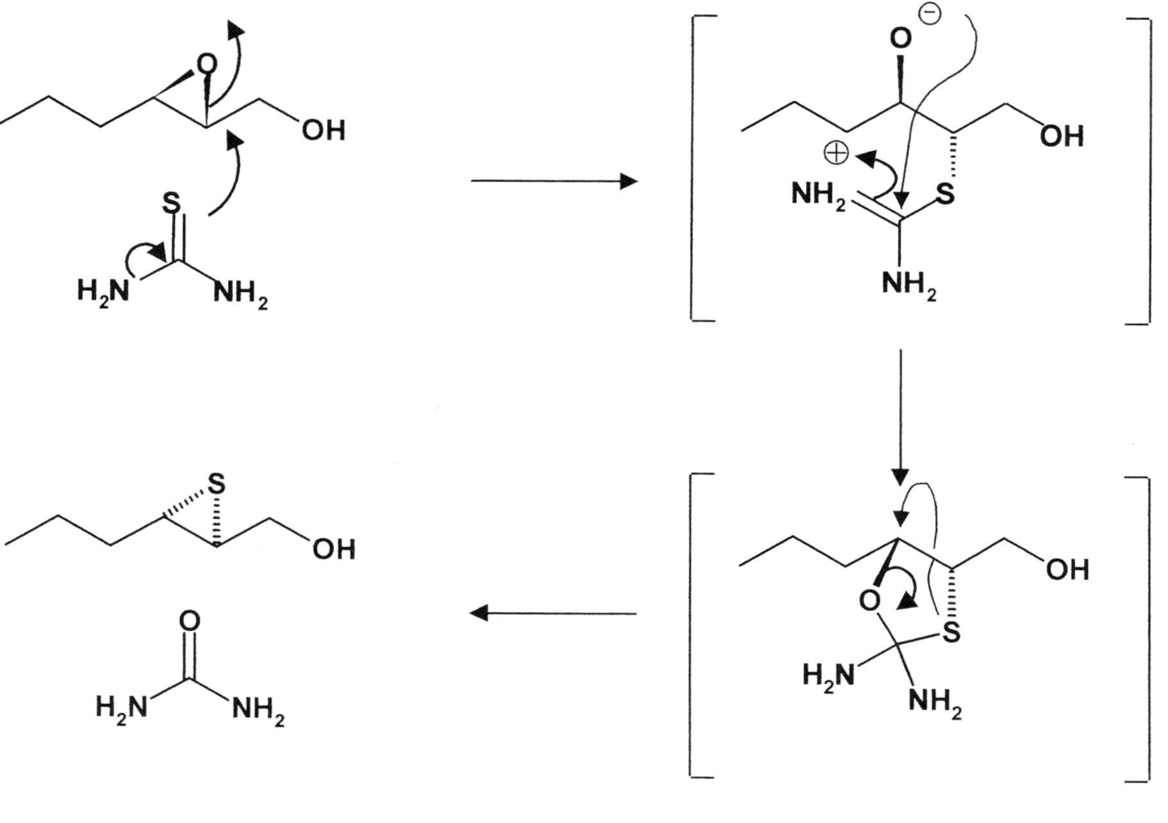

Scheme 3.

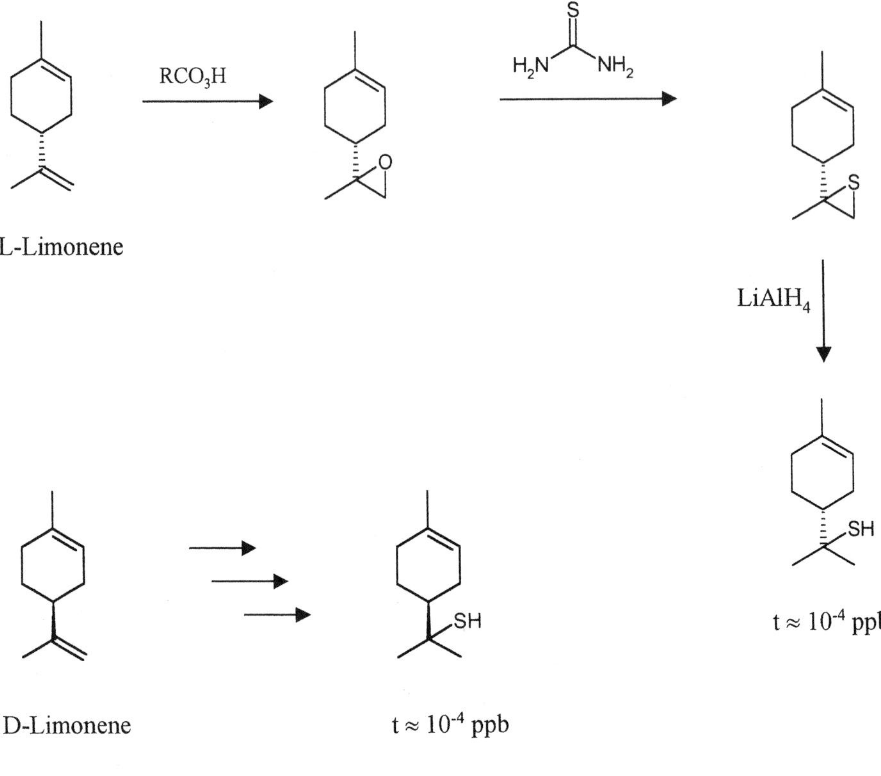

Scheme 4.

CH$_3$
OH
SH

5

Figure 3.

(2R,3S)-enantiomer

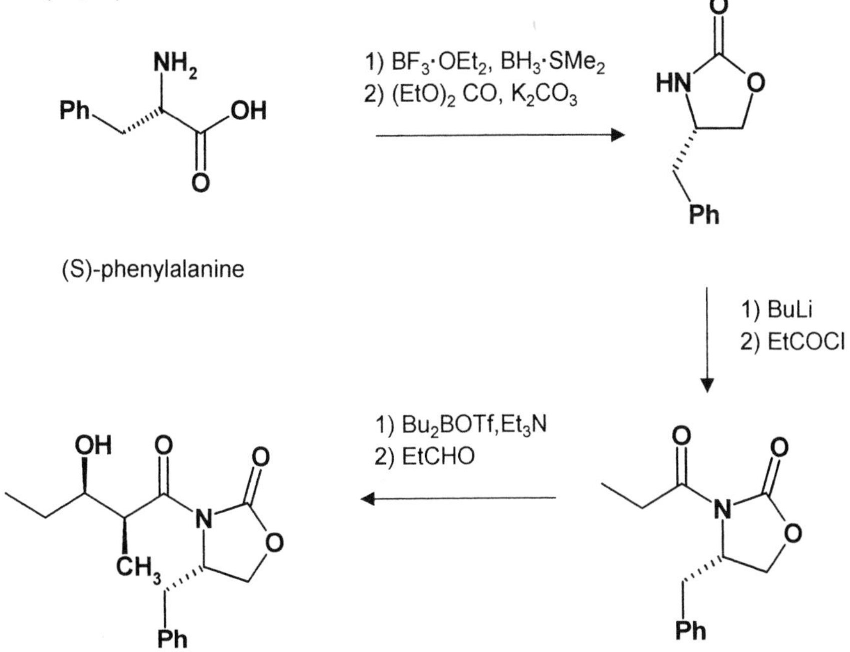

NH$_2$

Ph OH

O

(S)-phenylalanine

1) $BF_3 \cdot OEt_2$, $BH_3 \cdot SMe_2$
2) $(EtO)_2$ CO, K_2CO_3

HN O O Ph

1) BuLi
2) EtCOCl

1) Bu_2BOTf, Et_3N
2) EtCHO

OH O O N O CH$_3$ Ph

O O N O Ph

Scheme 5

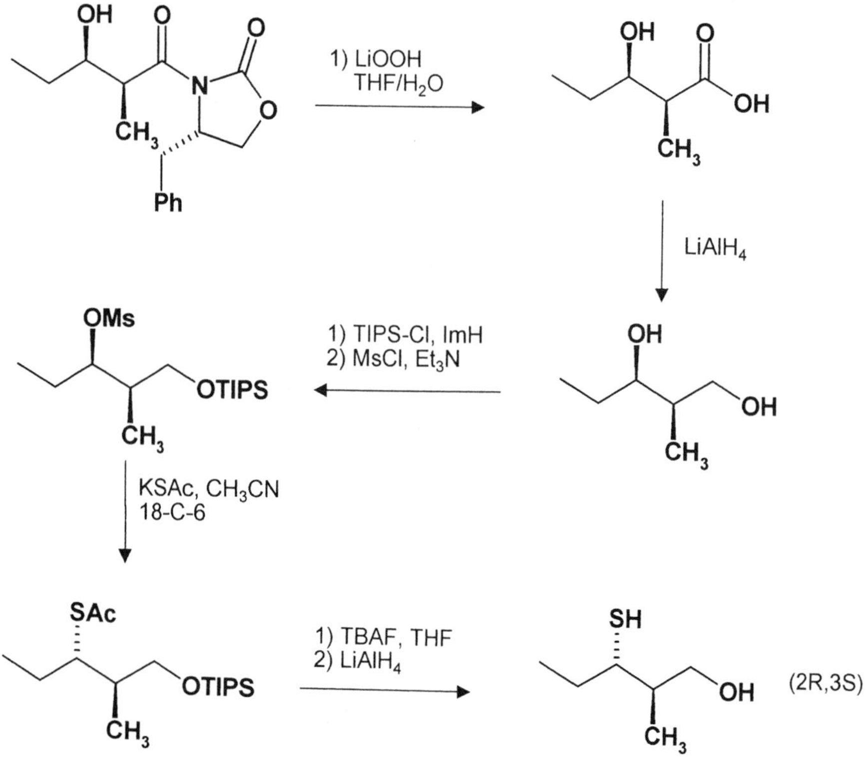
OH
O
O
N
O
CH3
Ph
1) LiOOH
THF/H2O
OH
O
OH
CH3
LiAlH4
OMs
OTIPS
CH3
1) TIPS-Cl, ImH
2) MsCl, Et3N
OH
OH
CH3
KSAc, CH3CN
18-C-6
SAc
OTIPS
CH3
1) TBAF, THF
2) LiAlH4
SH
OH
CH3
(2R,3S)

6

Scheme 5a

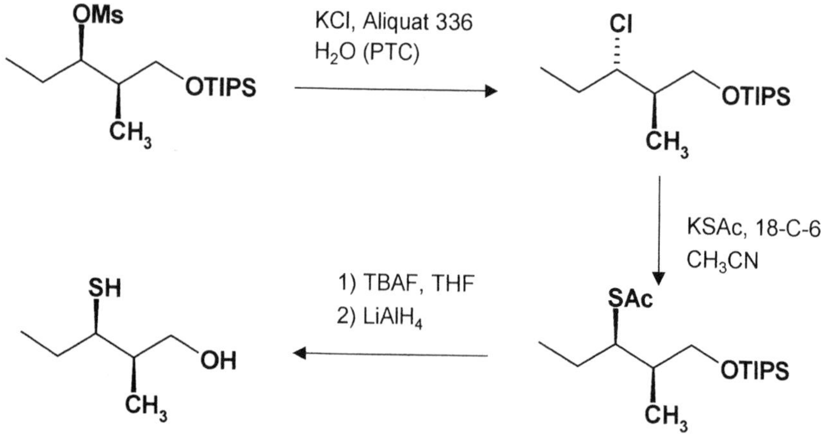
OMs
OTIPS
CH3
KCl, Aliquat 336
H2O (PTC)
Cl
OTIPS
CH3
KSAc, 18-C-6
CH3CN
SAc
OTIPS
CH3
1) TBAF, THF
2) LiAlH4
SH
OH
CH3

Scheme 5b Formation of the Syn-Product by Double Inversion of Configuration

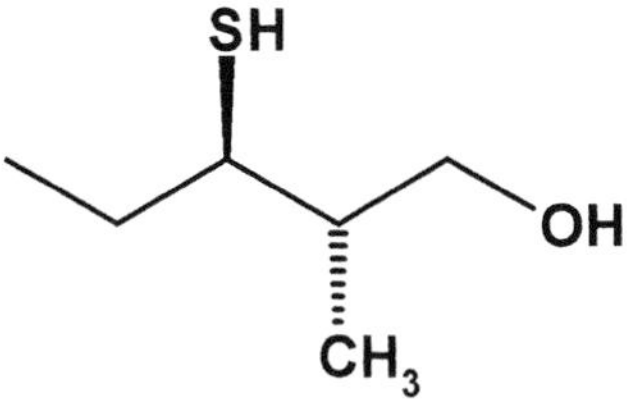

SH OH CH_3 SH OH CH_3

6 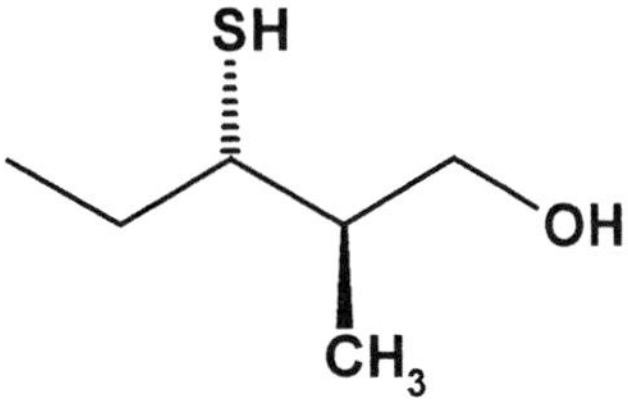 7

Figure 4.

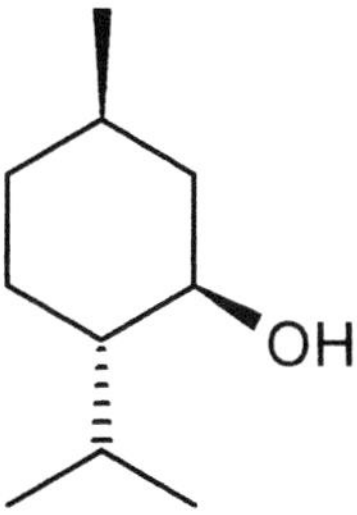

OH

8

Figure 5.

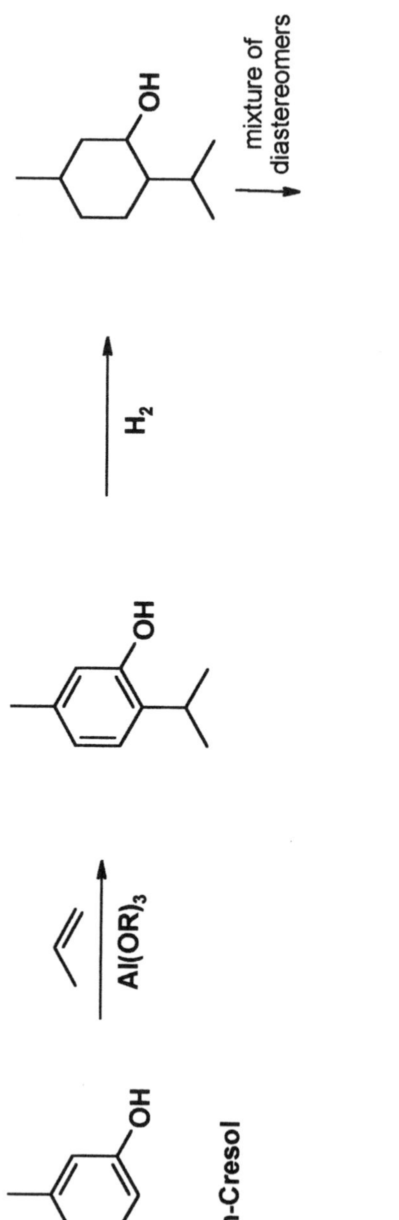

OH
m-Cresol
Al(OR)3
OH
H2
OH
mixture of
diastereomers
racemic menthol (+/-)
racemic menthyl benzoate (+/-)

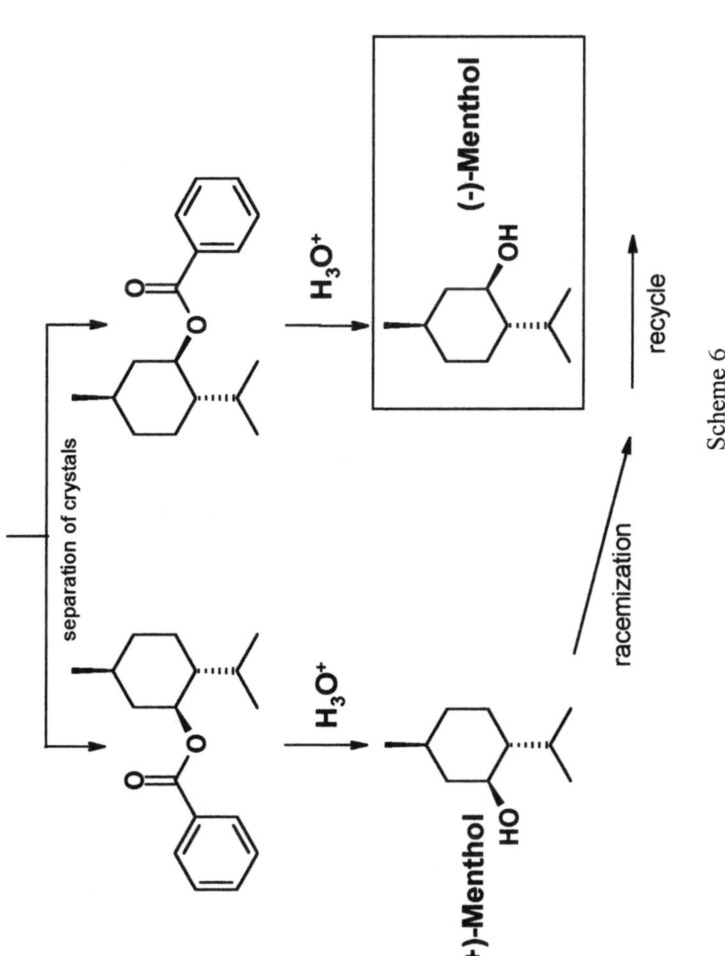
separation of crystals
O
O
O
O
H_3O^+
H_3O^+
(+)-Menthol
HO
(-)-Menthol
OH
racemization
recycle

Scheme 6

Use of Enantioselective Catalysts:

The Takasago synthesis for **8** uses the isomerization of diethyl-3,7-dimethyl-octa-2,6-dienylamine, made from myrcene and diethylamine, to the Schiff base of (R)-citronellal using S-BINAP-Rh® as catalyst for the induction of chirality (*10*) (scheme 7).

Another example is Fehr's enantioselective synthesis of the two forms of α-damascone **9** and **10**.

In this synthetic scheme the chirality is induced by an enantioselective protonation of an enolate with N-isopropyl-ephedrine as proton donor (*11,12*) (scheme 8).

The authors showed that the odor (S)(-)-α-damascone **9** is about 65 times stronger than that of its (R)(+)-enantiomer **10**.

Use Of Enzymes

Enantioselective hydrolyses of esters using enzymes from different natural sources are well-known procedures. Mori's enantioselective synthesis of (S)(-)-α-damascone **9** uses pig liver esterase to selectively hydrolyze 2,4,4-trimethylcyclohex-2-enylacetate to the (R)-alcohol which is then transformed to **9** according to scheme 9 (*13*).

Conclusion

A great number of the more than 7000 aroma chemicals known today to occur in edible materials (*14*) contain chiral centers in their molecular structure. The rapid increase of procedures to induce asymmetry during syntheses will make the enantiomeric forms of these molecules available to determine their sensory properties, which can be quite different, qualitatively, i.e. hedonically, and quantitatively, i.e. odor strength.

The cited examples are a selection of these procedures and their application to the enantioselective syntheses of some of these compounds, that in most cases show considerable differences in their sensory properties.

Further research in this field will make the more interesting enantiomeric forms available on an industrial scale as shown by the two l-menthol syntheses (*10*). This will contribute to create flavors that are even closer to the natural examples than they are today. Knowledge of the absolute configuration of these molecules, correlated to their sensory properties will also help to improve our knowledge about the mechanism of the interaction of odor stimulae with their receptors. (*15*).

References

1. Buck, L.; Axel, R. *Cell* **1991**, *65*, 175-187.

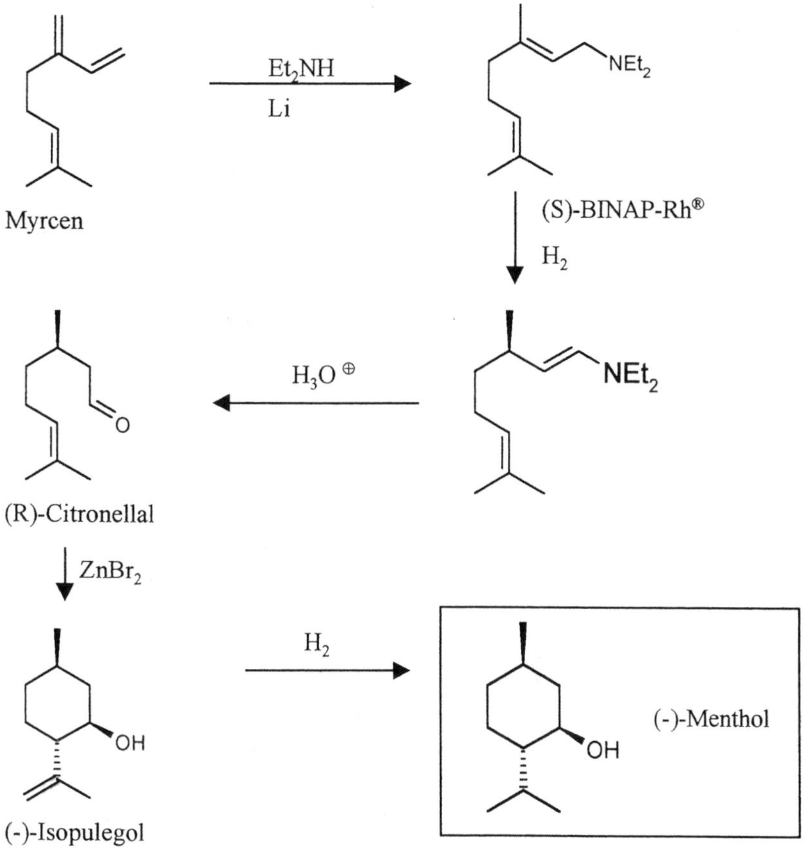

Myrcen

Et_2NH
Li

NEt_2

(S)-BINAP-Rh®
H_2

NEt_2

$H_3O^{\oplus}$

O

(R)-Citronellal

$ZnBr_2$

OH

(-)-Isopulegol

H_2

OH

(-)-Menthol

8

Scheme 7

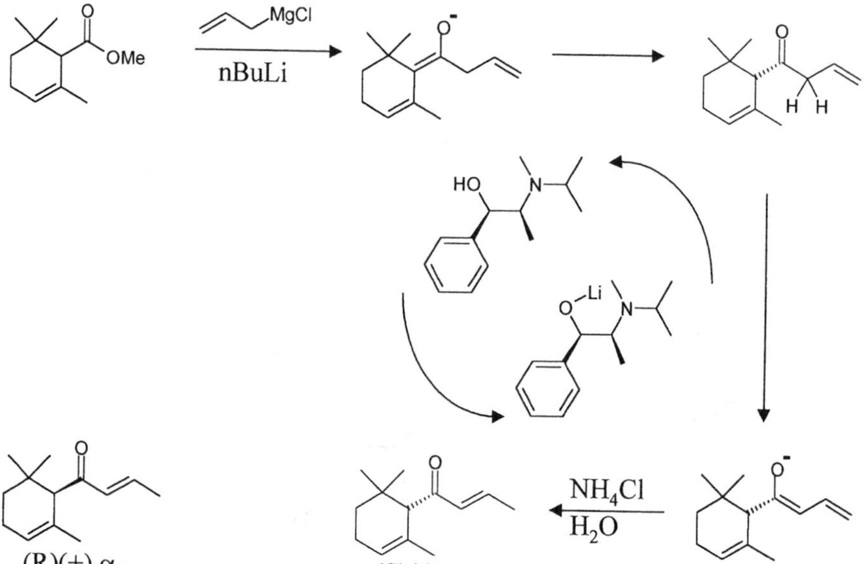

MgCl

OMe

nBuLi

HO

N

O

Li

H H

NH_4Cl

H_2O

(R)(+) α-
t = 100 ppb

10

(S)(-) α-
t = 1.5 ppb

9

Scheme 8

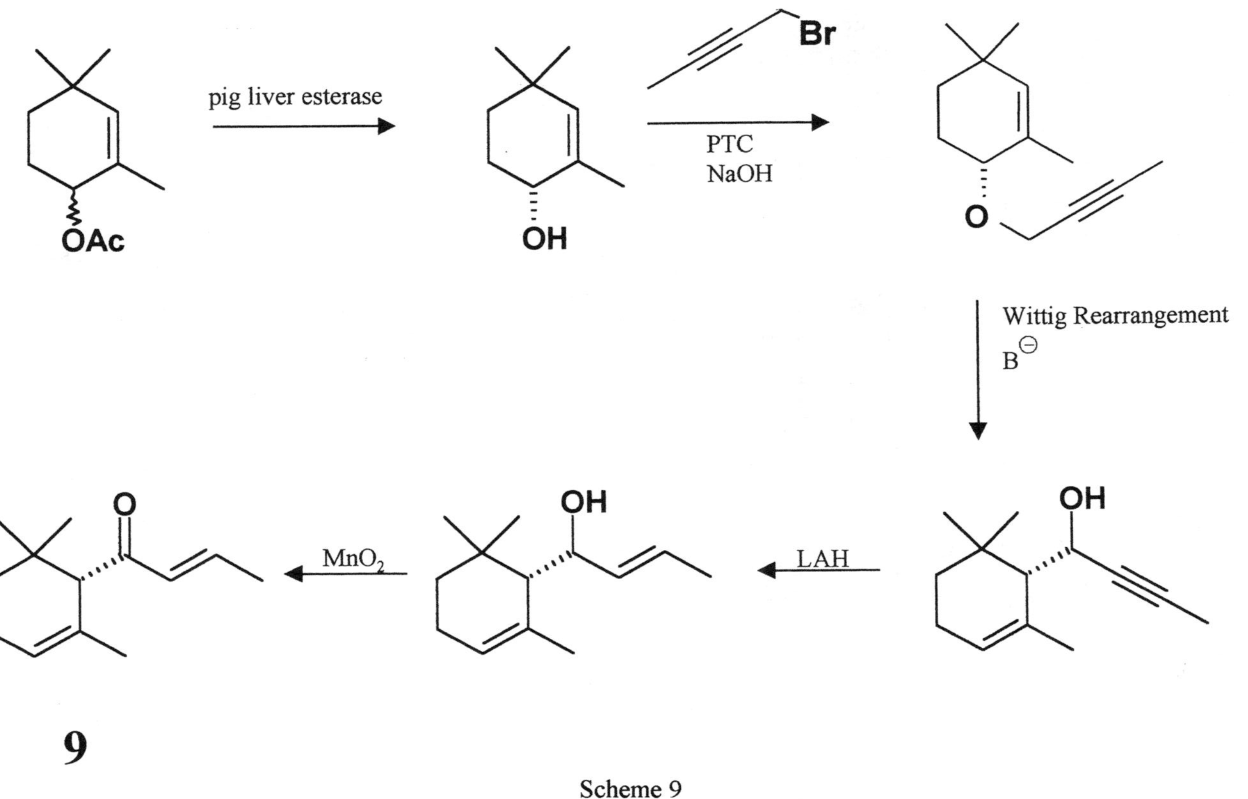
OAc
pig liver esterase
OH
Br
PTC
NaOH
O
Wittig Rearrangement
B⊖
OH
LAH
OH
MnO₂
O
9

Scheme 9

2. Pickenhagen, W. in *Flavor Chemistry: Trends and Developments*; Teranishi, R. ; Buttery, R.G.; Shahidi, F., Eds.; *ACS Symposium Series* 388; American Chemical Society: Washington, DC, 1989, pp 151-157.
3. Winter, M.; Furrer, A.; Willhalm, B.; Thommen, W. *Helv. Chim. Acta* **1976**, *59*, 1613-1620.
4. Pickenhagen, W.; Brönner-Schindler, H. *Helv. Chim. Acta* **1984**, *67*, 947-952.
5. Katsuki, T.; Sharpless, K.B. *J. Am. Chem. Soc.* **1980**, *102*, 5974-5976.
6. Demole, E.P.; Enggist, P.; Ohloff, G. *Helv. Chim. Acta* **1982**, *65*, 1785-1794.
7. Widder, S.; Sabater-Lüntzel, Ch.; Dittner, T.; Pickenhagen, W. *J. Agric. Food Chem.*, in press, 2000.
8. Sabater-Lüntzel, Ch.; Widder, S.; Vössing, T.; Pickenhagen, W. *J. Agric. Food Chem.*, in press, 2000.
9. Evans, D.A.; Bartroli, J.; Shih, T.L. *J. Am. Chem. Soc.* **1981**, *103*, 2127-2129.
10. Hopp, R., In *Actes des 14èmes Journées Internationales Huiles Essentielles, Digne-les-Bains, 31. August – 2. September 1995,* Rivista Italiana Eppos, 1996, pp 110-130.
11. Fehr, C.; Galindo, J. *J. Org. Chem.* **1988**, *53*, 1828-1830.
12. Fehr, C.; Galindo, J. *Angew. Chemie* **1994**, *106*, 1967.
13. Mori, K.; Amaike, A.; Hou, M. *Tetrahedron* **1993**, *49*, 1871-1878.
14. *Volatile Compounds in Food, Qualitative and Quantitative Data*, 7th Edition, 1996, including all Supplements and revisions until Autumn 1999, published by TNO Nutrition and Food Research, Zeist/Netherlands.
15. Wetzel, C.H.; Oles, M.; Wellerdieck, Ch.; Kuczkowiak, M.; Gisselmann, G.; Hatt, H. *J. of Neuroscience*, 1999, *19* (17), 7426-7433.

Chapter 13

Specificity of Glycosidases from Tea Leaves toward Glycidic Tea Aroma Precursors

Akio Kobayashi, Kikue Kubota, Dongmei Wang, and Takako Yoshimura

Laboratory of Food Chemistry, Department of Nutrition and Food Sciences, Ochanomizu University, 2–1–1 Ohtsuka, Bunkyo-ku, Tokyo 112–8610, Japan

A glycosidase mixture was separated and purified from fresh tea leaves. Its activity was determined through hydrolysis of *p*-nitrophenyl glucoside. Model mixtures with the same molar concentrations of glucosides and primeverosides (containing typical tea aroma compounds as the aglycons) were enzymatically hydrolyzed. The amounts of liberated volatile aglycons were quantitatively measured at regular time intervals. Primeverosides were hydrolyzed more rapidly than the glucosides. Different yields of individual glycosides were observed after 1 or 2 hours of incubation for the glucosides and after 15-60 min for the primeverosides, thus demonstrating the importance of primeverosidase in forming black tea aroma. However, glycosidase activity decreased throughout the black tea manufacturing process. The highest production of aroma compounds durimg black tea processing was due to interaction between the glycosides and the glycosidases.

Introduction

Among the three different types of tea, i.e., green, oolong and black, black tea is manufactured by a fermentation process which gives the characteristic color and aroma of this tea. Therefore, black tea is called fermented tea, although this epithet does not imply microbiological fermentation, but rather an endogeneous enzymatic action. The presence of glucose-bound aroma precursors was first pointed out by Saijo and Takeo (*1*), because the concentration of black tea aroma compounds was increased after maceration or by dilute acid hydrolysis of fresh tea leaves. However, the participation of glycosidase (*2*) in forming the black tea aroma compounds and the presence of hexenyl (*3*) and monoterpenyl (*4*) glucosides in fresh tea leaves were clarified 18 years later by the present authors. Since then, almost all of the black tea aroma compounds, i.e., (*Z*)-3-hexenol, benzyl alcohol, 2-phenylethanol, linalool, linalool oxides I, II, III and IV, geraniol and methyl salicylate, have been found as aglycons of mono- and disaccharides, i.e., glucosides, primeverosides, vicianosides and acuminosides. Among these saccharides, primeverosides were the most variable, and primeverosidase has been purified as the main glycosidase in fresh tea leaves (*5*). However, this enzymatic study focused on the aroma formation of oolong tea, and there are significant differences in the aroma constituents of black tea and oolong tea due to the different processing. We have qualitatively and quantitatively determined the glycosides in fresh tea leaves by GC and GC-MS analyses of their trifluoroacetylated (TFA) derivatives. The change in glycoside content throughout the black tea manufacturing process was also evaluated by the same method. The more rapid hydrolysis of primeverosides than that of glucosides resulted in the formation of various types of monoterpene alcohols, and primeverosidase is thought to be the main aroma formation enzyme during black tea manufacturing (*6*).

The present study looks in more detail at the enzymatic activity of a typical glycoside by using partially purified enzymes prepared from fresh tea leaves and discusses the discrepancy between the enzymatic activity and practical decrease in aroma precursors at various stages of the black tea manufacturing process.

Experimental

Separation and Purification of Glycosidases

Acetone powder from fresh tea leaves (*Camellia sinensis* var. *sinensis* cv. Yabukita), prepared according to Yano et al. (*2*), was dissolved in 0.1 M citrate buffer solution (pH 5.0) and treated as the crude enzyme solution. The crude enzyme fraction was precipitated by adding acetone up to 50% and again dissolved in the citrate buffer solution. The enzyme fraction was separated by adding ammonium sulfate to give 40-80% saturation. The precipitate was dissolved in the citrate buffer solution, dialyzed and centrifuged at low temperature to give a purified mixture of glycosidases in the supernatant. This procedure is summarized in Fig. 1.

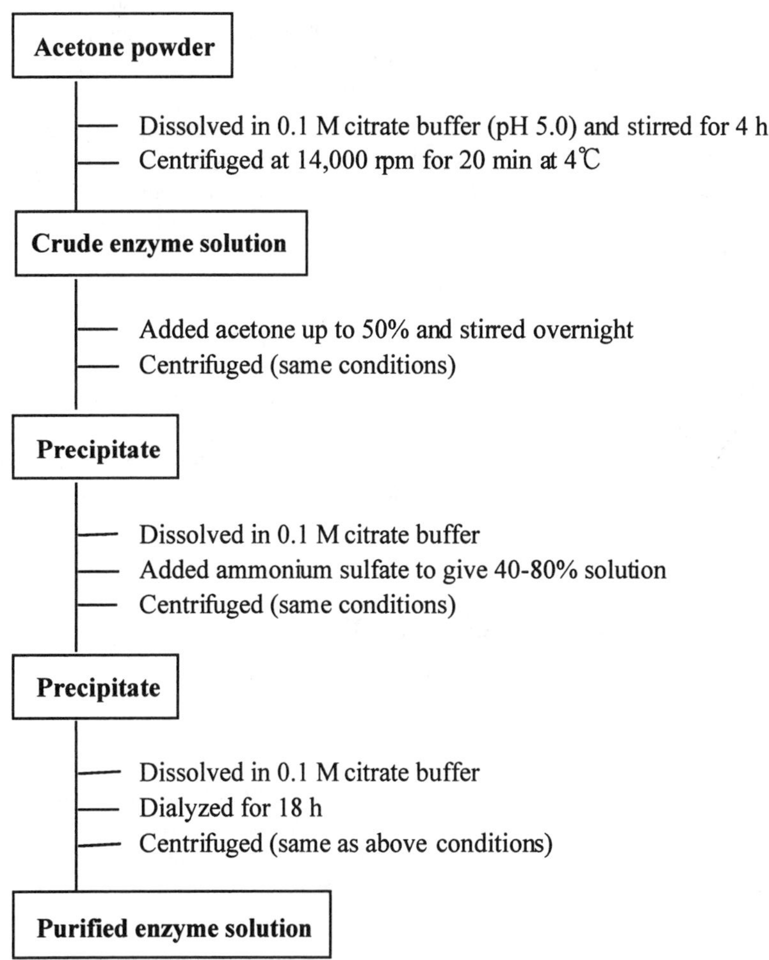

Figure 1. Purification of glycosidases.

Enzyme Assay

The *β*-D-glucosidase activity was determined by spectrometric measurement at 420 nm of *p*-nitrophenol (*p*NP) that was liberated by the hydrolysis of *p*NP-*β*-D-glucoside. The accurate quantity of *p*NP was determined from a calibration curve.

The amounts of liberated aglycons were determined by GC and GC-MS from the ratio of each to an internal standard after solvent extraction of the reacted solution. GC was conducted with a Hewlett Packard 5890 gas chromatograph equipped with a DB-WAX 60m X 0.25mm i.d. fused silica capillary column for general analyses and with a CP-Cyclodextrin-B-236-M-19 50m X 0.25mm i.d. fused silica capillary column for chiral compounds. The oven temperature was programmed from 100°C (for 10 min) to 200°C at 2°C/min. FID and MS-TIC were applied for detection.

Reagent and Reference Samples

Analytical grade solvents were further purified by distillation before use. *p*NP-*β*-D-glucopyranoside was purchased from Sigma. Benzyl-, (Z)-3-hexenyl-, geranyl-, 2-phenylethyl-, linalyl- and methyl salicylate-*β*-D-glucosides and their *β*-primeverosides and *p*NP-*β*-primeveroside were synthesized in our laboratory by a method described previously (*7*).

Results and Discussion

Purification of the Glycosidases

Since our purpose was to clarify the practical formation of black tea aroma by enzymatic hydrolysis, we did not try to separate each of the glycosidases or to determine individual glycosidase activity. Thus a partially purified enzyme mixture was used as the model for black tea aroma formation.

Enzyme Assay

Glycosidase activity was determined from the protein content and amount of hydrolyzed *p*NP from *p*NP-glucoside. One unit is defined as the amount of enzyme which hydrolyzed one µmol of substrate per minute at 37°C. Specific activity is expressed as units per milligram of enzyme protein. As shown in Table I, the specific activity of the purified solution was 25 times higher than that of the crude enzyme solution.

Table I. Hydrolytic Activities of the Crude and Purified Enzyme Solutions Toward *p*NP-glucoside.

	Protein Content	Activity	Specific Activity	Purification Fold
Crude enzyme	89.7 (mg/g of AP)	2.1 (unit/g of AP)	0.02 (unit/mg protein)	1.0
Purified enzyme	0.6 (mg/ml of sol.)	0.3 (unit/ml of sol.)	0.5 (unit/mg protein)	25.0

Specificity to the Aglycon and Saccharide Moieties:

The tea glucosides bearing five tea aroma compounds as aglycons were mixed at the same molecular concentration (50 μl of each 16 mM sample) with a two fold concentration of racemic linalool glucoside, The resulting mixture was preincubated with 23 ng of nonanol as an internal standard at 37°C for 5 min. The purified enzyme solution (500 μl) was then added and incubation was continued under the same conditions. After 0.5, 1, 2, 6 and 24 h respectively, aliquots of the solution were heated to stop the enzyme activity. The liberated aglycons were extracted with diethyl ether. The concentrated extracts were analyzed by GC and GC-MS and the amount of each aglycon was calculated from its peak area ratio to that of nonanol (i.s.). This procedure is shown by flow chart in Fig. 2.

Since the total time for black tea processing takes about 24 h, we selected the concentrations of the glycosides to be exhausted after 24 h. The time-course plots of the yield of each volatile compound are shown in Fig. 3. The yields of liberated aglycons from glucosides varied in the earliest stage up to 30 min. This trend continued through 2 h and up to 6 h of incubation, at which time methyl salicylate, one of the optically active linalool isomers, 2-phenylethanol and geraniol showed the same highest yield. From 6 h to 24 h, the increase in yield of each sample moderated and all yields except for that of benzyl alcohol reached 10 times that of the internal standard. At this point ≥90% of the glucosides were estimated to be hydrolyzed.

The hydrolysis rate of benzyl glucoside was exceptionally low, thus explaining the low concentration of benzyl alcohol among the tea aroma constituents in contrast to the predominance of benzyl glucoside among the glucosides in fresh tea leaves. A different hydrolytic ratio was observed for *R*- and *S*-linalyl glucoside. The yield was analyzed by GC using a modified *β*-cyclodextrin capillary column. *R*-Linalool always showed a higher yield compared to *S*-linalool, and the glucosidase seemed to be specific with regard to the stereochemistry of the aglycon. However, this difference was not apparent in the hydrolysis of linalyl primeveroside by the same enzyme solution.

The experimental conditions we selected were the same as those of the

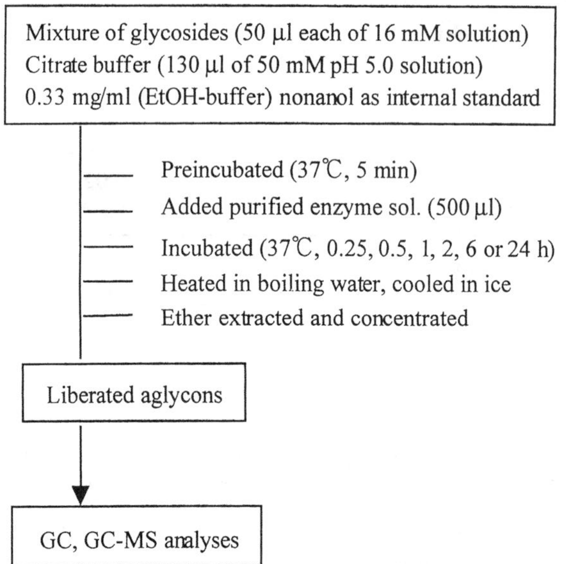

Figure 2. Quantitative analytical procedure for hydrolyzed aglycons from glycosides.

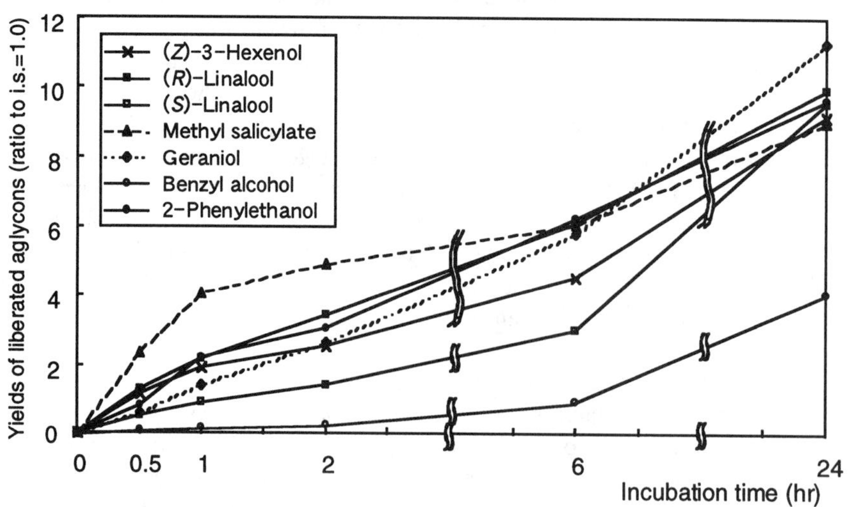

Figure 3. Hydrolysis of the synthesized glucosides.

primeverosides were not clearly distinguishable with the yield of aglycons.

Changes in Enzymatic Activity during Manufacturing

The results just described prompted us to try to correlate the enzymatic activity and the decrease in glycosidic aroma precursors.

The glycosides in tea leaves decrease during the manufacturing process, namely the plucking (of fresh leaves), withering, rolling and fermentation stages. The glucosides decrease less than the primeverosides, and we postulate that primeverosidase is the main enzyme forming the characteristic black tea aroma. By following the separation method already described, we prepared four crude enzyme solutions from the four stages of processing and compared their enzymatic activities. The crude enzyme solutions were prepared under the conditions described in the experimental section. The glycosidase activity of each was measured by using *p*NP-glucoside and *p*NP-primeveroside as substrates.

Fig. 5 demonstrates that the decreasing amounts of corresponding glucosides and primeverosides were consistent with the changes in the two glycosidase activities. Unexpectedly, the activity decreased throughout the process and, at the rolling stage, the primeverosidase activity was less than that of glucosidase. The highest activity at the fresh tea leaf stage does not mean the highest production of aroma at this stage, because the substrates can not interact with the enzymes. During the process while hydrolyzing the substrates, the enzymes are also being manufacturing process. However, the reaction times showing the variable yields of volatiles differed between the incubation experiment and tea processing. Two hours of incubation were enough to obtain different but reasonable yields of the glucosidase derived volatiles, compareed to 7-8 h which are necessary to complete the rolling and fermentation processes. This discrepancy in reaction time between the model system and the manufacturing process can be explained by the heterogeneous interaction between glycosides and glycosidase probably requiring a longer time to complete the fermentation in the manufacturing process.

The same tendency was apparent when the primeverosides were used as substrates (Fig 4). The different yield of enzymatically liberated aglycons was observed from 15 min to 2.0 h of incubation and may explain the higher primeverosidase activity compared to other glycosidases in fresh tea leaves as we have stated previously (*7*). Benzyl primeveroside was more rapidly hydrolyzed than the glucoside. These results are in accordance with the different decreases of these two glycosides by the hydrolytic tea enzymes in our previous experiments (*6*).

We have shown in the preceding report that the primeverosides in fresh tea leaves decreased considerably during the rolling and fermentation stages. The levels of 2-phenylethyl-, geranyl- and methyl salicylate primeverosides decreased more rapidly than those of the benzyl- and (*Z*)-3-hexenyl primeverosides. Fig. 4 shows that methyl salicylate primeveroside was more efficiently hydrolyzed than (*Z*)-3-hexenyl primeveroside, particularly during 0 min to 30 min incubation. The other

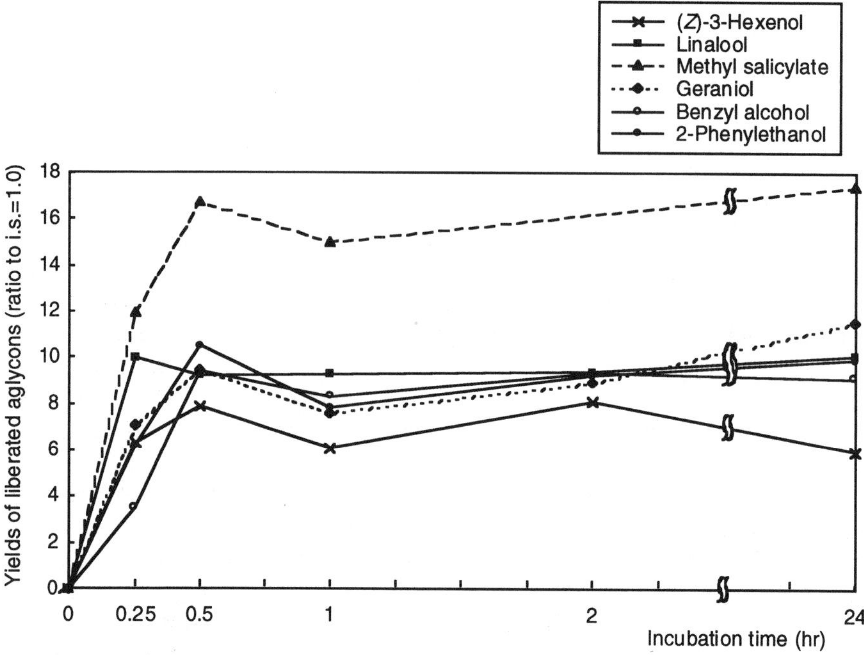

Figure 4. Hydrolysis of the synthesized primeverosides.

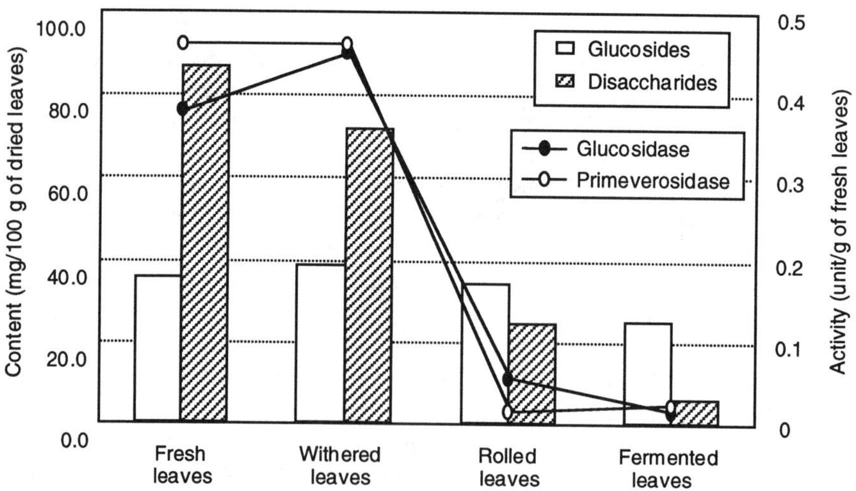

Figure 5. Changes in glycoside contents and enzyme activities during black tea processing.

deactivated. The primeverosidase activity after the rolling stage is low, however, the hydrolysis of primeverosides was most active at this stage. We are now wondering whether *p*NP-primeverosidase activity is suitable for determining the total primeverosidase activity, because this enzyme is known to be specific to the aglycon moiety.

Conclusion

1) A purified enzyme solution was prepared from fresh tea leaves, and its glycosidase activity was determined.
2) (*Z*)-3-Hexenyl-, linalyl-, methyl salicylate-, geranyl-, benzyl- and 2-phenylethyl-glucosides and primeverosides were hydrolyzed by this enzyme solution, and the yields of their aglycons were determined. Specific yields of the aglycons were observed within 1-2 h for the glucosides and within 15 min - 1 h for the primeverosides.
3) The glycosidase activity decreased throughout the black tea manufacturing process. The highest production of aroma compounds during fermentation was not due to activation of the enzymes, but to interaction between the substrates and enzymes.

References

1. Saijo, R.; Takeo, T. *Agric. Biol. Chem.* **1972,** *37*, 1367-1373.
2. Yano, M.; Okada, K.; Kubota, K.; Kobayashi, A. *Agric. Biol. Chem.* **1990,** *54*, 1023-1028.
3. Kobayashi, A.; Kubota, K.; Koke, Y.; Wada, E. *Biosci. Biotech. Biochem.* **1994,** *58*, 592-593.
4. Morita, K.; Wakabayashi, M.; Kubota, K.; Kobayashi, A.; Herath, N. *Biosci. Biotech. Biochem.* **1994,** *58*, 687-690.
5. Ogawa, K.; Iijima, Y.; Guo, W.; Watanabe, N.; Usui, T; Dong, S.; Tong, Q.; Sakata, K. *J. Agric. Food Chem.* **1997**, *45*, 877-882.
6. Kobayashi, A.; Kubota, K.; Wang, D. 9th Weurman Flavour Research Symposium, Freising, Germany, 21-25 June 1999; p.18 (due to be published by the year of 2000).
7. Matsumura, S.; Takahashi, S.; Nishikitani, M.; Kubota, K.; Kobayashi, A. *J. Agric. Food Chem.* **1997**, *45*, 2674-2678.

Chapter 14

Biosynthesis of γ-Nonalactone in Yeast

L.-A. Garbe, H. Lange, and R. Tressl

Institut für Biotechnologie, Technische Universität Berlin, Seestrasse 13, D–13353 Berlin, Germany

γ-Nonalactone is known as aroma active compound in fermented products such as beer. In a series of labeling experiments with deuterated linoleic acid and [$^{18}O_1$]-13- and 9-hydroxyoctadecadienoic acid, respectively, two biosynthetic routes to γ-nonalactone in yeast were elucidated: (i) 13-lipoxygenation / reduction and β-oxidation followed by one α-oxidation step results in (*S*)-γ-nonalactone (~ 60 % e.e.); (ii) 9-lipoxygenation / reduction and Baeyer-Villiger oxidation yields azelaic acid and 2*E*,4*E*-nonadien-1-ol which is further transformed into (*R*)-γ-nonalactone (~ 46 % e.e.).

Introduction

γ-Lactones are widely distributed important flavor compounds occurring in fruits as well as in fermented foods such as beer. The biosynthesis of γ-decalactone and γ-dodecalactone has been investigated in fruits and microorganisms *(1-3)*. The common initial step is the introduction of oxygen into the carbon chain. In *Sporobolomyces odorus* metabolization of oleic acid proceeds via (*R*)-12-hydroxylation and subsequent degradation of the resultant ricinoleic acid by β-oxidation yielding (*R*)-γ-decalactone *(2)*. Epoxygenation of oleic or palmitoleic acid has been characterized as additional pathway operative in *S. odorus (3)*. Metabolization of the intermediate epoxy fatty acid leads to γ-dodecalactone and γ-decalactone, respectively. The stereochemical course of the microbial degradation of the epoxy- and their corresponding dihydroxy fatty acids is currently under investigation *(4)*.

Degradation of these oxygenated fatty acids by β-oxidation yields lactones with an even number of carbon atoms. The formation of lactones with odd numbered chain lengths, e.g. γ-nonalactone, cannot be explained by this pathway. Furthermore, γ-nonalactone obtained by yeast-catalyzed reactions exhibits low optical purity compared to γ- or δ-decalactone (Table I).

Experimental

Synthesis of [9,10,12,13-$^{2}H_4$]-9,12-octadecadienoic acid and its lipoxygenase products (hydroperoxyoctadecadienoic acid, (HPOD) and hydroxy-octadecadienoic acid (HOD)) were performed as described previously *(5)*. [9- or 13-$^{18}O_2$]-9- or 13-HPOD and [9- or 13-$^{18}O_1$]-9- or 13-HOD were synthesized by enzymatic lipoxygenation of linoleic acid with tomato or soy bean lipoxygenase under [$^{18}O_2$]-oxygen gas (HPOD) and subsequent reduction with $NaBH_4$ (HOD), respectively.

Incubation experiments (150 or 250 mg/L of substrate) were performed in 1000 mL shake flasks with 200 ml yeast cultures. *Saccharomyces cerevisiae* (IfG 06136, RH-strain) was purchased from the IfGB Berlin, Germany. *Sporobolomyces odorus* (CBS 2636) was obtained from the CBS, The Netherlands.

Table I. Configuration and Optical Purity of Naturally Occurring γ-Nonalactone

Source	*Configuration*	*Enantiomeric purity % e.e.*
S. odorus ATCC 24259	(R)	44
S. odorus ATCC 26697	(R)	66
milk	(R)	80
Cheddar cheese	(S)	13

Consequently, we investigated the biosynthesis of γ-nonalactone in *Saccharomyces cerevisiae* and *Sporobolomyces odorus* in order to elucidate the enzymatic steps involved in the formation of lactones with odd numbered carbon chains starting from linoleic acid.

Results and Discussion

In a series of labeling experiments two different pathways were shown to be involved in the biosynthesis of γ-nonalactone in yeast (Figure 1).

Pathway I is based on a 13-lipoxygenase / reductase catalyzed route involving subsequent β-oxidation of 13-HOD into (*S*)-δ-decalactone. In isotopic labeling experiments [13-$^{18}O_{1(2)}$]-(*S*)-13-H(P)OD as well as [9,10,12,13-$^{2}H_2$]-(*S*)-13-H(P)OD and [13-$^{2}H_1$]-(±)-13-HOD were transformed into labeled (*S*)-δ-decalactone (~ 60-100% e.e.).

(*S*)-δ-Decalactone and (*S*)-5-hydroxydecanoic acid are further metabolized in *S. cerevisiae* by α-oxidation, a new metabolic route shown to be very effective for metabolization of (*S*)-configurated 5- and 3-hydroxy fatty acids in yeasts.

α-Oxidation pathways are known in plants and were recently found to be involved in the metabolization of 3-methyl branched fatty acids in mammalian liver tissue. The yeast *S. cerevisiae* obviously uses an enzymatic α-oxidation step to transform (*S*)-5-hydroxydecanoic acid into (*S*)-4-hydroxynonanoic acid and (*S*)-γ-nonalactone (Figure 2). The origin of γ-nonalactone by this pathway could be demonstrated by labeling experiments with ^{2}H-, ^{18}O- and ^{13}C- precursors.

γ-Nonalactone and 4-hydroxynonanoic acid are further metabolized in *S. cerevisiae* by α-oxidation. The resultant 3-hydroxyoctanoic acid shows (*S*)-configuration with high enantiomeric purity (~ 96 % e.e.). This is in accordance with a metabolization of (*R*)-3-hydroxyoctanoic acid by the multifunctional protein (MFP) of yeast possessing (*R*)-stereospecificity *(6)*. (*S*)-3-hydroxy-octanoic acid is transformed to hexanoic acid by two α-oxidation steps.

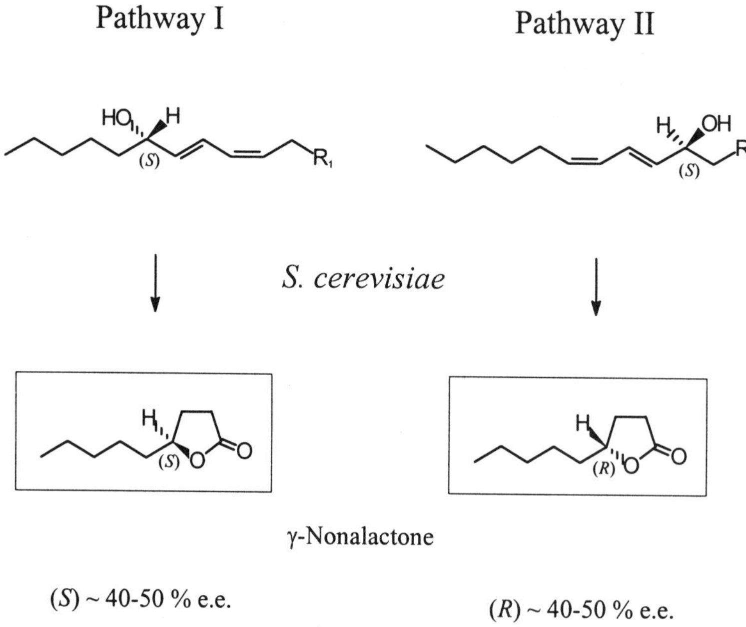

Figure 1. Biosynthetic routes to γ-nonalactone in yeast (R = -(CH$_2$)$_6$-COOH).

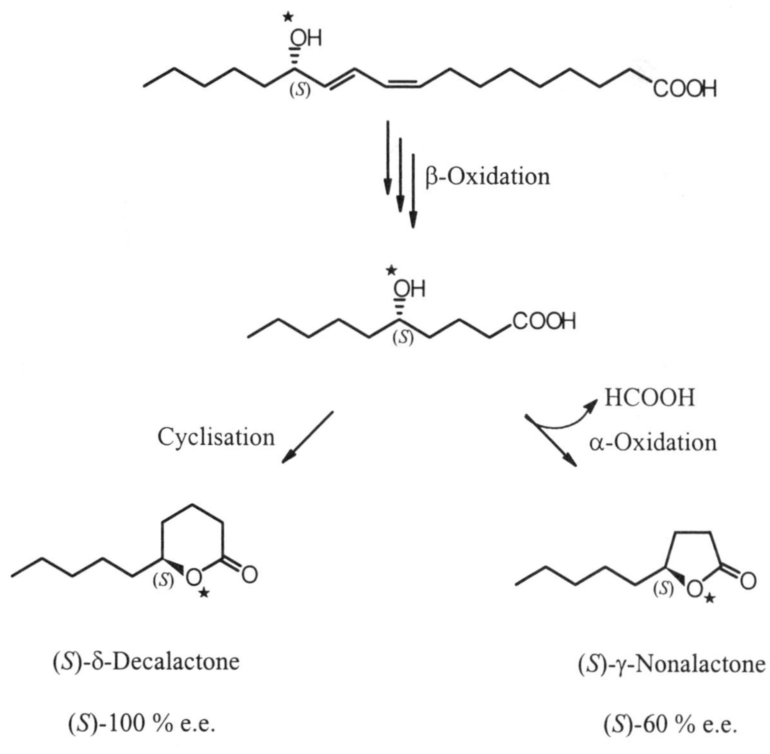

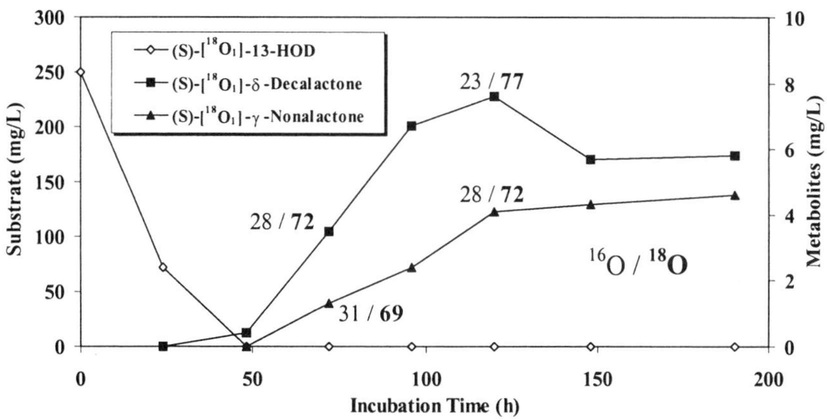

Figure 2. Metabolization of [13-$^{18}O_1$]-(S)-13-Hydroxy-9Z,11E-octadecadienoic acid (150 mg/L) by S. cerevisiae.

The (*S*)-configuration of the C_{18}-hydro(pero)xyde initially formed by lipoxygenation is in accordance with the (*S*)-configuration of the formed δ-decalactone (92 % e.e.). γ-Nonalactone resulting from this pathway also shows (*S*)- configuration (50-70 % e.e.).

The alternative pathway II to γ-nonalactone is based on a 9-lipoxygenase / reductase catalyzed reaction of linoleic acid. Linoleic acid, [9,10,12,13-$^2H_4$]-(*S*)-9-hydro(pero)xy-10*E*,12*Z*-octadecadienoic acid as well as [9-$^2H_1$]-(±)-9-hydroxy-10*E*,12*Z*-octadecadienoic acid are converted to 9-oxo-10*E*,12*Z*-octadecadienoic acid, isomerized to 9-oxo-11*E*,13*E*-octadecadienoic acid and oxygenated by a Baeyer-Villiger enzyme. The formed Baeyer-Villiger ester is hydrolyzed to yield azelaic acid and 2*E*,4*E*-nonadien-1-ol. 2*E*,4*E*-Nonadien-1-ol is further oxidized to 2*E*,4*E*-nonadienoic acid, transformed to 3*Z*-nonenoic acid by acyl-CoA-reducto-isomerase and metabolized within the "epoxyde pathway" of yeast to yield (*R*)-γ-nonalactone (Figure 3). This "epoxyde pathway" has been intensively studied in the yeast *S. odorus (3)*. Incubation experiments using deuterated 9-hydro(pero)xy-10*E*,12*Z*-octadecadienoic acids yielded γ-nonalactone (~50 % e.e.).

Conclusion

The studies revealed two completely different pathways to be operative in the degradation of 13- and 9-hydro(pero)xyoctadecadienoic acids in yeast. A 13-lipoxygenase pathway with α-oxidation activity and a 9-lipoxygenase pathway with enzymatic Baeyer-Villiger oxidation activity were characterized in yeast cleaving the carbon chain of oxygenated linoleic acid into C_9 fragments which are further transformed into γ-nonalactone and metabolized by peroxysomal β-oxidation, respectively.

Acknowledgment

We are grateful to „Arbeitsgemeinschaft industrieller Forschungsvereinigungen (AiF), Otto von Guericke e. V.“ for financial support of this study in the course of the research project No. 11428 N. This project has been supported by funds of „Bundesminister für Wirtschaft (BMWi)“.

References

1. Schöttler, M.; Bohland, W. *Helv. Chim. Acta* **1996**, *79*, 1488-1496.
2. Haffner, T.; Tressl, R. *J. Agric. Food Chem.* **1996**, *44*, 1218-1223.

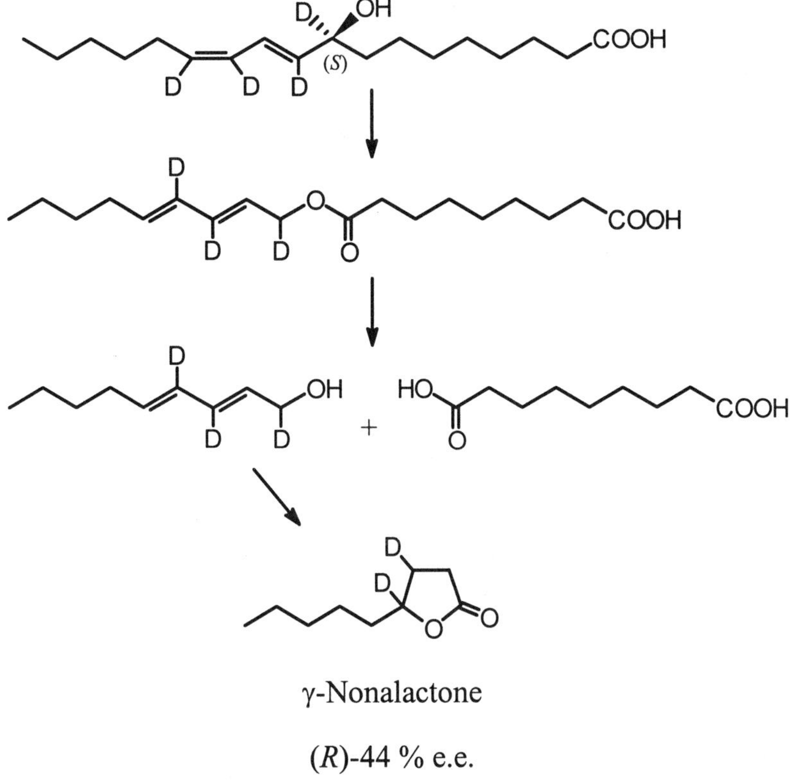

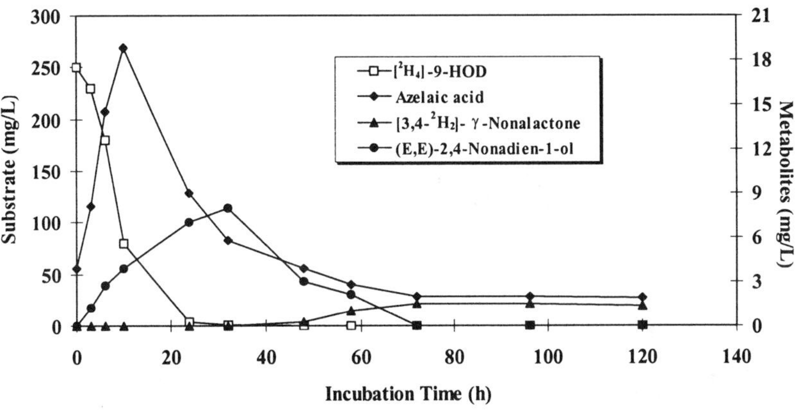

Figure 3. Metabolization of [9,10,12,13-$^2H_4$]-(S)-9-Hydroxy-10E,12Z-octadecadienoic acid (250 mg/L) by S. odorus.

3. Haffner, T.; Tressl, R. *Lipids* **1998**, *33*, 47-58.
4. Garbe, L.-A.; Tressl, R, unpublished
5. Albrecht, W.; Schwarz, M.; Heidlas, J.; Tressl, R. *J. Org. Chem.* **1992**, *57*, 1954-1956.
6. Filppula, S.; Sormunen, R. T.; Hertig, A.; Kunau, W. H.; Hiltunen, K. *J. Biol. Chem.* **1995**, *270*, 27453-27457.

Chapter 15

The Formation of Strecker Aldehydes

H. Weenen and J. G. M. van der Ven

Bio-Organic Chemistry Section, Quest International, P.O. Box 2, 1400 CA Bussum, The Netherlands

Strecker degradation is an important reaction for flavour formation during the heating of foodstuffs, as well as in the preparation of process flavours. To study the formation of aldehydes in Strecker degradation, phenylalanine was reacted with various co-reactants, including aldoses, ketoses, Amadori rearrangement products, and various α-dicarbonyl containing compounds such as C_4, C_5, and C_6 3-deoxyglycosones. The reaction was followed by measuring the formation of the Strecker aldehyde during Likens-Nickerson distillation-extraction. Of all co-reactants measured in our study, pyruvaldehyde was most effective in generating Strecker aldehydes. Among the aldoses tested, erythrose was the best phenylacetaldehyde precursor, which is consistent with the observation that 3-deoxyerythrosone was the most efficient Strecker aldehyde co-reactant of all tested 3-deoxyosones. Finally, the Strecker aldehyde generating ability was studied of the ARP's of glucose and valine, of glucose and proline, and of glyceraldehyde and proline.

Introduction

Strecker degradation is a key reaction in the thermal treatment of foods, and in process flavour formation (1, 2). Strecker degradation results a.o. in the formation of CO_2 and Strecker aldehydes (Figure 1) (3, 4). Many Strecker aldehydes are important flavour substances (Table 1), *e.g.* methional, isobutanal, 2-methylbutanal and 3-methylbutanal. Strecker aldehydes can also be incorporated in other flavour

Table I. Amino acids, their corresponding Strecker aldehydes, and organoleptic properties

No	*Amino acid*	*Strecker aldehyde*	*Organoleptic properties*	*Odor threshold (ppb)*	*Ref.*
1	glycine	formaldehyde	mouse urine, ester-like	50000	10
2	alanine	acetaldehyde	pungent, fruity	25	10
3	valine	2-methylpropanal	pungent, fruity, chocolate	0.1 - 2.3 / 2	10, 11
4	leucine	3-methylbutanal	fruity, peach, cocoa, malty	0.2 / 3	10, 12
5	iso-leucine	2-methylbutanal	cocoa, fresh, fruity	1 / 4	10, 12
6	serine	hydroxy-acetaldehyde	n.a.[1]	n.a.	
7	threonine	2-hydroxypropanal	n.a.	n.a.	
8	phenylal anine	phenylacetaldehyde	honey like, sweet, flowery	4	11
9	tyrosine	4-hydroxyphenyl-acetaldehyde	n.a.	n.a.	
10	methioni ne	methional	cooked potato	0.2	12
11	proline	1-pyrroline / 4-aminobutanal	spermous	22	13
12	ornithine	1-pyrroline / 4-aminobutanal	spermous	22	13
13	cysteine	mercapto-acetaldehyde	sweet, sulfury, rotten fruit[2]	n.a.	14

[1] n.a. = not available
[2] Oganoleptic properties of the dimer.

substances *e.g.* thiazoles, oxazoles, thianes, thiolanes, and alkylpyrazines (5-8). Strecker degradation of amino acids and α-dicarbonyl containing substances not only gives CO_2 and aldehydes, but also generates α-aminocarbonyls which are pyrazine precursors (Figure 2) (6, 9). Amino acids that generate ammonia can form pyrazines by mechanisms other than Strecker degradation as well (6).

Strecker degradation of some amino acids e.g. proline and cysteine, gives rise to unstable aldehydes (respectively figures 3 and 4), however, they often react further to more stable products with important flavour properties (5, 15-17). Amino acids that generate relatively stable aldehydes include glycine, alanine, valine, leucine, isoleucine, methionine, phenylalanine, and tyrosine.

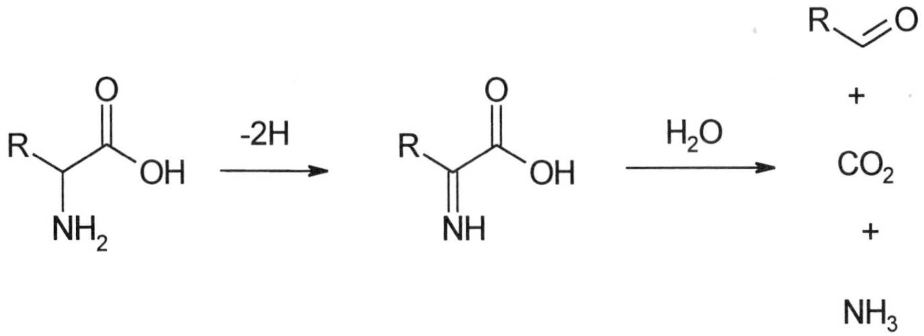

Figure 1. General equation of the Strecker degradation.

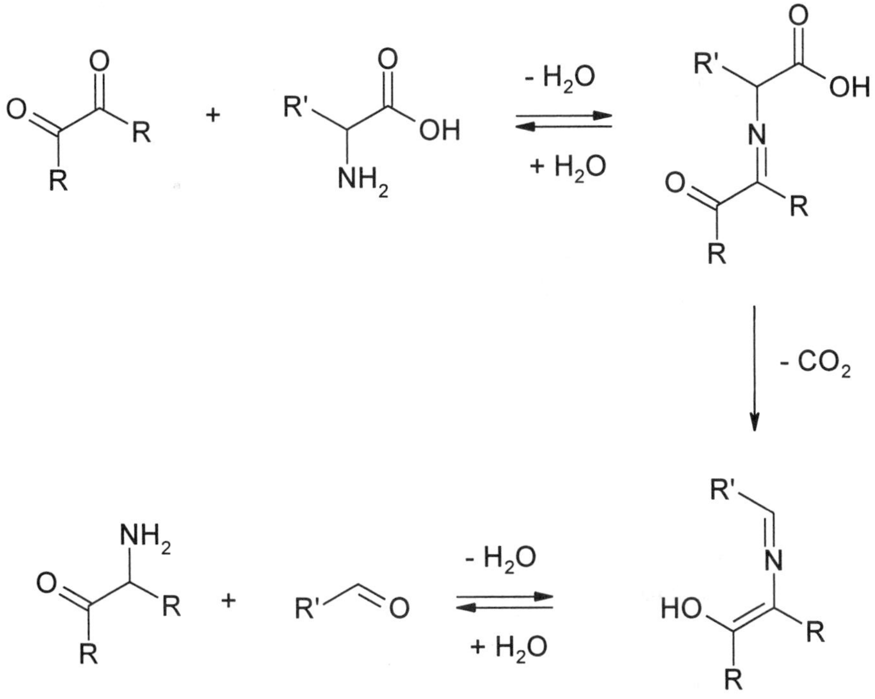

Figure 2. Strecker degradation of an α-amino acid and an α-dicarbonyl.

Strecker degradation has been known since 1862, when A. Strecker first described that an α-amino acid can react with a carbonyl compound in aqueous solution or suspension to give CO_2 and an aldehyde containing one carbon atom less (18). Reviews on the reaction scope and mechanism have been reported (3, 9).

Since only few and inconclusive studies have been reported on the role of Maillard reaction intermediates in Strecker degradation (19), we have investigated Strecker degradation of phenylalanine when reacted with monosaccharides, 3-deoxyosones, as well as some smaller carbohydrate fragments, and ARP's.

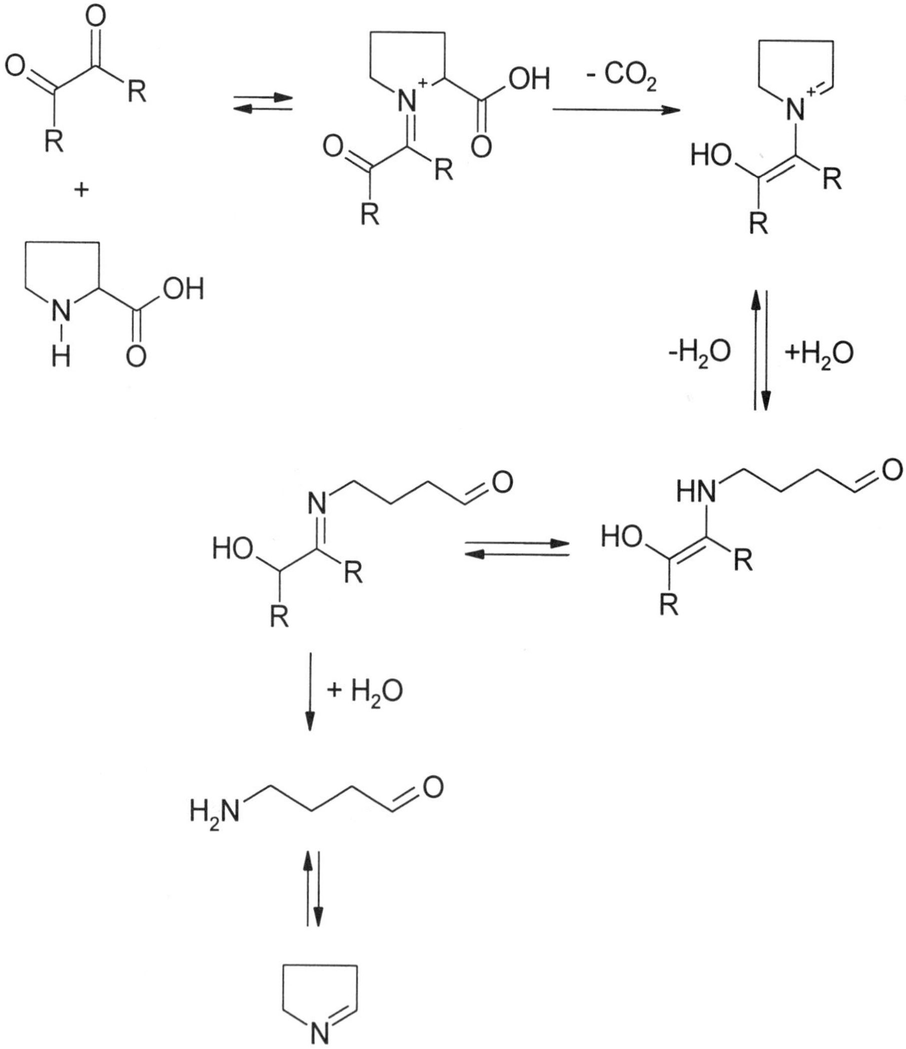

Figure 3. Strecker degradation of proline.

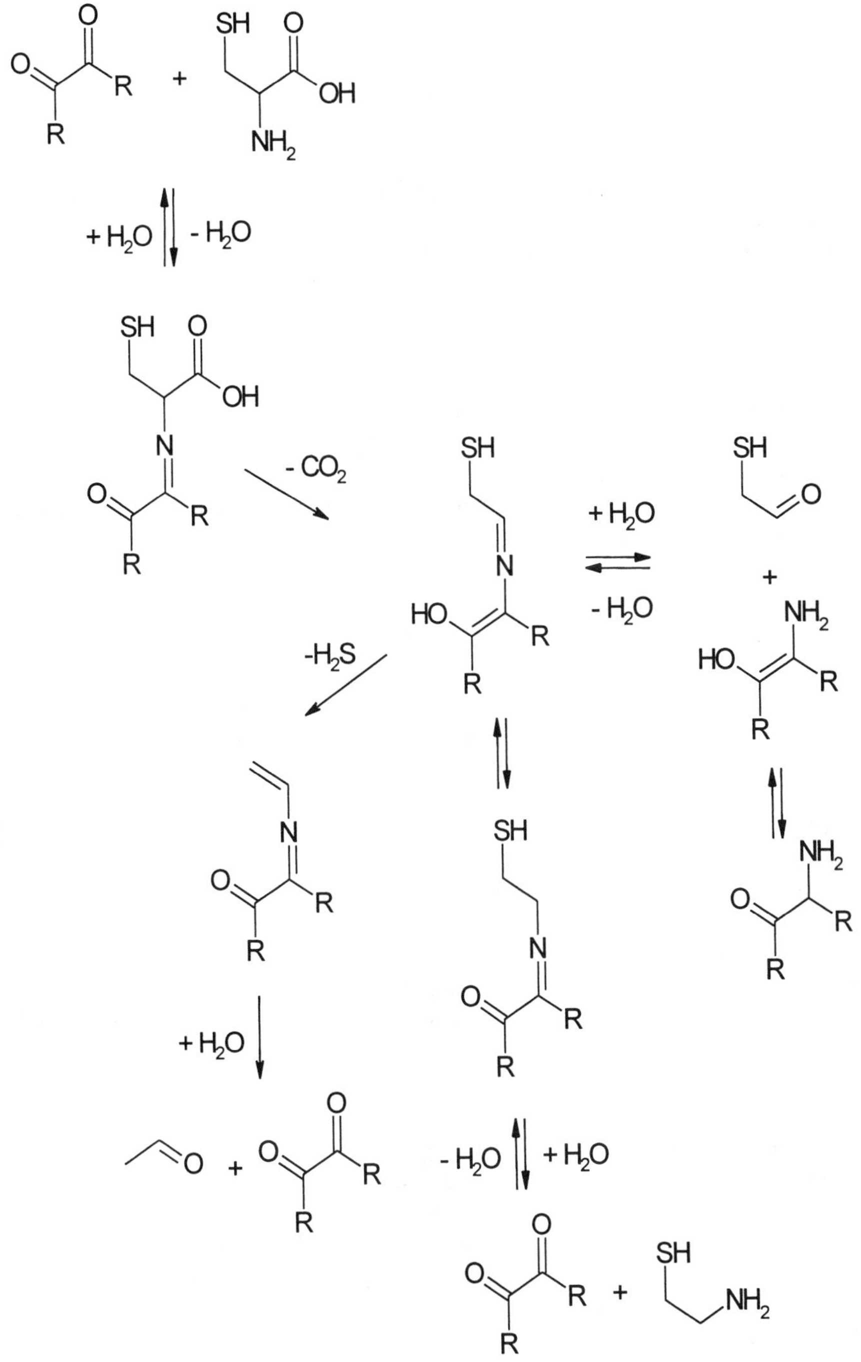

Figure 4. Strecker degradation of cysteine.

Experimental

Synthesis

The ARP of glucose and proline was synthesized according to Vernin et al. (20), the ARP of glyceraldehyde and proline according to Huyghues-Despointes and Yaylayan (21), the ARP of glucose and phenylalanine according to Sosnovsky et al. (22), and the ARP of glucose and valine according to Xenakis et al. (23).

Analysis

GC was performed using a Carlo Erba HRGC 5300 gas chromatograph equipped with a flame ionization detector (FID) and a HP-5 column (50 m × 0.32 mm i.d.; 1.05 μm film). The column temperature was programmed to rise from 75 to 150 °C at a rate of 3 °C /min, followed by an increase to 300 °C at a rate of 40 °C/min, and maintained at that temperature for 5 min. The injector and the detector temperatures were 125 and 210 °C, respectively.

Degradation experiments

Degradation experiments with proline or proline ARP's as reactant were performed using a Likens-Nickerson steam distillation extraction apparatus (24). The ARP (150 μmol-2.7 mmol), or the parent carbohydrate and amino acid in the same molar concentrations, were dissolved in phosphate buffer (35 mL, 50 mM, pH 7.2) or citrate buffer (50 mL, 50 mM, pH 3.5), and continuously extracted for 4 h with dichloromethane to which 2-acetylpyrazine was added as the internal standard. Quantitation was based on comparison with external standard consisting of solution of ACTP, ACP and acetylpyrazine. Concentration – response correlation was linear in the concentration range in which measurements took place. After 4 h, heating of the flask containing the aqueous phase was stopped and extraction with dichloromethane continued for another 20 min. The organic phase was analysed for 2-acetyl-3,4,5,6-tetrahydropyridine/2-acetyl-1,4,5,6-tetrahydropyridine (ACTP) and 2-acetyl-1-pyrroline (ACP) using the GC method described above.

Phenylacetaldehyde experiments were performed using a Likens-Nickerson steam distillation extraction apparatus (24). The carbonyl containing compound (~ 1.3 mmol) and an equimolar amount of phenylalanine, were dissolved in citrate buffer (50 mL, 50 mM, pH 3.5), and continuously extracted for 2 h with dichloromethane to which undecane was added as the internal standard. After 2 h, heating of the flask containing the aqueous phase was stopped and extraction with dichloromethane continued for another 20 min. The organic phase was analysed for phenylacetaldehyde using the GC method described above.

For experiments in ethanol, a mixture of the ARP (~ 1 mmol), potassium dihydrogenphosphate (170 mg, 1.25 mmol), and potassium hydrogenphosphate trihydrate (285 mg, 1.25 mmol) in ethanol was heated under reflux. After 2 h, the mixture was quickly cooled down and a sample was mixed with a stock solution of 2-acetylpyrazine in ethanol, as the internal standard, and the concentration of phenylacetaldehyde determined using the GC method described above.

Results

ARP as phenylacetaldehyde precursor

When heating an aqueous solution of the glucose-phenylalanine ARP, phenylacetaldehyde is generated. As is shown in table 2 formation of this Strecker aldehyde is enhanced if the ARP is used instead of a mixture of the corresponding carbohydrate and amino acid. Surprisingly, degradation of a phenylalanine solution without glucose results in the same amount of phenylacetaldehyde (Table 2, entry 4) as the glucose and phenylalanine mixture.

When studying the effect of the pH on the formation of phenylacetaldehyde it was found that for ARP's a low pH (3.2) is advantageous for the formation of the Strecker aldehyde, however, phenylacetaldehyde formation from glucose and phenylalanine is more efficient at pH 7. The decarboxylation rate in the reaction of alanine with glyoxal has also been reported to increase with increasing pH (19). Some care should be taken when comparing the results at pH 3.2 and 7.0, as different buffer salts were used at these pH's. Phosphate is known to be a catalyst for Maillard reactions, the mechanism of which has been described as general base catalysis (25). Since the citrate anion should have similar properties, the difference in activity between phosphate and citrate is not expected to be significant.

If the degradation of the ARP is performed in ethanol (Table 2, entry 5) only a very small amount of phenylacetaldehyde is formed, in addition to furfural.

Comparison of the volatiles formed during degradation of glucose and valine, with those formed during degradation of the corresponding ARP confirmed the superior performance of ARP's as Strecker aldehyde precursors. However, it will be shown below that this may not always be the case.

Carbohydrates and α-dicarbonyl containing species in the Strecker degradation of phenylalanine.

The influence of a number of α-dicarbonyls, α-hydroxycarbonyls and other reactive Maillard intermediates on the Strecker degradation of phenylalanine was studied in more detail. These included C_4, C_5, and C_6 3-deoxyglycosones, two glucose derived ARP's, pyruvaldehyde, and glyoxal. The results of these experiments are shown in table 3. For the series of aldoses tested erythrose is the best carbohydrate in generating phenylacetaldehyde. Ketoses (dihydroxyacetone and fructose) were better than the corresponding aldoses (glyceraldehyde respectively glucose) in generating phenylacetaldehyde. Of the two ketoses tested, dihydroxyacetone generated the largest amount of phenylacetaldehyde. Among the reactive Maillard intermediates, there is a clear increase in phenylacetaldehyde formation going from C-6 3-deoxyosone to the C-3 3-deoxyosone, *i.e.* pyruvaldehyde. The latter compound is the best dicarbonyl compound for Strecker aldehyde formation in our study, which is in agreement with earlier reports in which some of the reactants addressed in our study were investigated (26).

Table II. Formation of phenylacetaldehyde from the glucose-phenylalanine ARP, during a Likens-Nickerson steam distillation-extraction procedure.

Entry	*Starting materials*	*Conditions*	*pH*	*Yield of phenylacetal dehyde*
1	Glucose + Phenylalanine	phosphate buffer (100 mM), reflux, 2 h	7.0	0.15%
2	ARP from glucose and phenylalanine	phosphate buffer (100 mM), reflux, 2 h	7.0	0.63%
3	Glucose + Phenylalanine	citrate buffer (50 mM), reflux, 2 h	3.2	0.006%
4	Phenylalanine	citrate buffer (50 mM), reflux, 2 h	3.2	0.006%
5	ARP from glucose and phenylalanine	citrate buffer (50 mM), reflux, 2 h	3.2	1.6%
6	ARP from glucose and phenylalanine	EtOH, phosphate salts (100 mM), reflux, 2h	7.0	very small amount, also furfural formed

There is a clear benefit in using compounds that are more advanced in the Maillard reaction: 3-deoxyglucosone is better in generating phenylacetaldehyde, followed by the ARP of glucose and phenylalanine, which is better than glucose. The ARP of glucose and proline is more a precursor of the 1-deoxyglucosone than of 3-deoxyglucosone. 1-Deoxyglucosone is much more reactive, which is in agreement with the larger amount of phenylacetaldehyde formed from the ARP of glucose and proline.

α-Hydroxycarbonyls are also good precursors of Strecker aldehydes (Table 3). This can be explained by their ability to generate α-dicarbonyls through cleavage (monosaccharides, deoxyosones), elimination (monosaccharides, glyceraldehyde, dihydroxyacetone) or after hydrolysis (deoxyosones). Notable exceptions to this hypothesis however, are hydroxyacetone and glycolaldehyde. Although they are less efficient than respectively dihydroxyacetone and glyceraldehyde as Strecker aldehyde precursors, they are clearly more efficient than glucose, fructose, and even xylose. This indicates that at pH 3.5 α-hydroxycarbonyls can efficiently react with amino acids to give Strecker aldehydes, without α-dicarbonyl involvement, suggesting an H_2O elimination mechanism.

Table III. Formation of phenylacetaldehyde at pH 3.2.

Carbonyl compound	*Yield of phenylacetaldehyde*	*Carbonyl compound*	*Yield of phenylacetaldehyde*
Pyruvaldehyde	21.0 %	Glc-Pro ARP	0.9 %
3-deoxyerythrosone	17.2 %	Glycolaldehyde	0.8 %
Dihydroxyacetone	9.0 %	3-deoxyglucosone	0.6 %
Glyoxal	8.8 %	Xylose	0.6 %
Erythrose	5.7 %	Glc-Phe ARP*	0.5 %
Glyceraldehyde	3.7 %	Fructose	0.13 %
3-deoxyxylosone	2.6 %	Glucose	0.03 %
Hydroxyacetone	1.2 %		

* Glc-Phe ARP was reacted without added phenylalanine.

ARP's as 1-pyrroline precursors.

1-Pyrroline, the expected Strecker degradation product of proline, is unstable and can therefore not be quantified reliably as such. Since the formation of 2-acetyl-1,4,5,6-tetrahydropyridine (ACTP) and 2-acetyl-1-pyrroline (ACP) is dependent on the formation of 1-pyrroline according to the proposed reaction mechanisms (15, 27), we assumed their formation as indicative for the ability of the proline ARP's to form 1-pyrroline.

In table 4 yields of the formation of ACP and ACTP are shown when the ARP's of glucose and proline, respectively glyceraldehyde and proline are degraded in a Likens-Nickerson apparatus at pH 7.2, and are compared with the formation of ACP and ACTP directly from the corresponding starting compounds. Clearly, there is no benefit in using these Maillard intermediates for the formation of ACP and ACTP. In fact in the case of ACTP formation from the glucose-proline ARP, the yield was more than 4 × as low when compared with a glucose and proline mixture!

Table IV. Formation of ACP and ACTP from Amadori rearrangement products.

Starting compound	*Yield ACP*	*Yield ACTP*
Glc-Pro ARP	0.016%	0.04%
Glc + Pro	0.017%	0.18%
Glyceraldehyde-Pro ARP	0.033%	0.24%
Glyceraldehyde + Pro	0.034%	0.29%

Discussion

Deoxyosones are α-dicarbonyl precursors (28, 29) and as such can be expected to induce Strecker degradation in combination with amino acids. Fujimaki et al. (27) describe that 3-deoxyglucosone has weak reactivity in Strecker degradation, when heated with L-leucine in distilled water at 80 °C for 30-60 min. Under these conditions 3-deoxyglucosone generated 560 × less isovaleraldehyde than pyruvaldehyde. On the other hand Ghiron et al. (30) conclude in their study of the reaction of 3-deoxyglucosone and phenylalanine (heated in distilled water at 100 °C for 30 min), that an important role of 3-deoxyglucosone is to participate in Strecker degradation resulting in the formation of aldehydes.

This study shows that deoxyosones are moderate to excellent Strecker degradation precursors, depending on the number of carbons in the molecule:

3-deoxyerythrosone > 3-deoxyxylosone > 3-deoxyglucosone.

In general reactivity in Strecker degradation seems correlated with how easily a reactant can form a (preferably small) free α-dicarbonyl substance. This is, however, not an absolute requirement, as hydroxyacetone and hydroxyacetaldehyde are reasonably good Strecker reactants. Among the 3-deoxyosones described above, the ease of free α-dicarbonyl formation seems to correlate well with their Strecker degradation precursor efficiency (31).

Glucose was the least effective Strecker aldehyde reactant of all compounds tested. Maillard reaction intermediates which result when glucose is exposed to Maillard reaction conditions, were all much more reactive than glucose itself:

3-deoxyglucosone > glucose-phenylalanine ARP > glucose.

3-Deoxyglucosone was not much more reactive than the ARP of glucose and phenylalanine, suggesting that further activation of these intermediates may result from fragmentation (32), which can take place from both the ARP or 3-deoxyglucosone. Interestingly the ability of the ARP of glucose and proline to produce phenylacetaldehyde from phenylalanine (via Strecker degradation) was almost twice as high as the ability of the ARP of glucose and phenylalanine to form phenylacetaldehyde (respectively 0.9 and 0.5 % yield). This may be due to the fact that secondary amines and amino acids generate mainly the more reactive 1-deoxyglucosone.

During the course of writing up these results, a study on Strecker degradation was presented by Th. Hofmann and P. Schieberle (Weurman symposium, Munich, 1999), which indicated that oxygen catalyzes the formation of carboxylic acids during Strecker degradation. Since all experiments described here were carried out under identical conditions, relative yields should not be significantlly affected by this phenomenon.

Acknowledgement

The technical assistance of D.-J. Treffers and K. Kruithof is gratefully appreciated.

Literature Cited

1. F. Ledl and E. Schleicher. New aspects of the Maillard reaction in foods and in the human body, *Angew. Chem. Int. Ed. Engl.*, 29, 1990, 565-594.
2. H. Weenen and J.F.M. de Rooij. Process flavourings, in: *Flavourings*, E. Ziegler and H Ziegler, Eds., Wiley-VCH, Weinheim, Germany, 1998, 233-258.
3. A. Schönberg and R. Moubacher. The Strecker degradation of α-amino-acids, *Chem. Rev.*, 50, 1952, 261-277.
4. G.P. Rizzi. The Strecker degradation and its contribution to food flavor, 1998, in press.
5. E.J. Mulders. Volatile components from the non-enzymatic browning reaction of the cysteine/cystine-ribose system, *Z. Lebensm. Unters. Forsch.*, 1973, 152, 193-201.
6. H. Weenen, S.B. Tjan, P.J. de Valois, N. Bouter, A. Pos, H. Vonk. Mechanism of pyrazine formation, in: *Thermally generated flavors*, ACS symposium series, ACS, Washington, D.C., 1994, 142-157.
7. M. Boelens, L.M. van der Linde, P.J. de Valois, H.M. van Dort, and H.J. Takken. Organic sulfur compounds from fatty aldehydes, hydrogen sulfide, thiols, and ammonia as flavour constituents, *J. Agric. Food Chem.*, 36, 1988, 677-680.
8. E.-M. Chui, M.-C. Kuo, L.J. Bruechert, and C.-T. Ho. Substitution of pyrazines by aldehydes in model systems, *J. Agric. Food Chem.*, 38, 1990, 58-61.

9. G.P. Rizzi. New aspects on the mechanism of pyrazine formation in the Strecker degradation of amino acids, in: *Flavor science and technology*, M. Martens, G. A. Dalen, H. Russwurm jr., John Wiley & Sons Ltd., 1987, 23-28.
10. Belitz, H.-D., Grosch, W. In: '*Lehrbuch der Lebensmittelchemie*', 3rd ed., Springer Verlag, Berlin, Germany.
11. Leffingwell, J.C. Flavour-base 1998, Leffingwell & Ass., Canton, GE, USA.
12. Guadagni, D.G., Buttery, R.G., Turnbough, J.G. Odour thresholds and similarity ratings of some potato chip components. *J. Sci. Food Agric*., 1972, 23, 1435-1444.
13. Amoore, J.E., Forrester, J., Buttery, R.G. Specific anosmia to 1-pyrroline: the spermous primary odour, *J. Chem. Ecol.,* 1, 299-310 (1975).
14. A. van Delft. Personal communication.
15. P. Schieberle, Quantitation of important roast-smelling odorants in popcorn by stable isotope dilution assays and model studies on flavor formation during popping, *J. Agric. Food Chem.*, 43, 1995, 2442-2448.
16. Y. Zheng, S. Brown, W.O. Ledig, C. Mussinan, and C.-T. Ho. Formation of sulfur-containing flavor compounds from reactions of furaneol and cysteine, glutathione, hydrogen sulfide, and alanine/sodium sulfide, *J. Agric. Food Chem*, 45, 1997, 894-897.
17. F. Chan and G.A. Reineccius. Kinetics of the formation of methional, dimethyl disulfide, and 2-acetylthiophene via the Maillard reaction, in: *Sulfur compounds in foods*, C.J. Mussinan and M.E. Keelan, Eds., ACS symposium series 564, ACS, Washington, D.C., 1994, 127-137.
18. A. Strecker. Notiz über eine eigenthümliche Oxydation durch Alloxan, *Ann. Chem.*, 123, 1862, 363-365.
19. N.V. Chuyen, T. Kurata, and M. Fujimaki. Studies on the Strecker degradation of alanine with glyoxal, Agric. Biol. Chem., 36, 1972, 11199-1207.
20. G. Vernin, L. Debrauwer, G.M.F. Vernin, R.-M. Zamkotsian, J. Metzger, J.L. Larice and C. Parkanyi. Heterocycles by thermal degradation of Amadori intermediates, in: *Off-flavors in foods and beverages*, G. Charalambous, Ed., Elsevier, Amsterdam, the Netherlands, 1992, 567-623.
21. A. Huyghues-Despointes and V.A. Yaylayan. Retro-aldol and redox reactions of Amadoricompounds: Mechanistic studies with variously labeled D-[13-C]Glucose, *J. Agric. Food Chem.*, 44, 1996, 672-681.
22. G. Sosnovsky, C.T. Gnewuch, and E.-S. Ryoo. Role of *N*-nitrosated Amadori compounds derived from glucose-amino acid conjugates in cancer promotion or inhibition, *J. Pharm. Sci.*, 82, 1993, 649-656.
23. D. Xenakis, N. Moll, and B. Gross. Organic synthesis of Amadori rearrangement products, *Synthesis*, 1983, 541-543.
24. G.B. Nickerson and S.T. Likens. Gas chromatographic evidence for the occurrence of hop oil components in beer, *J. Chromatog.*, 21, 1966, 1-5.
25. R.P. Potman, Th.A. van Wijk. Mechanistic studies of the Maillard reaction with emphasis on phosphate-mediated catalysis, in: *Thermal generation of aromas,* Th.H. Parliament, R.J. McGorrin, C-T Ho, Eds., ACS-symposium series 409, ACS, Washington, DC, 1989, 182-195.

26. M. Fujimaki, N. Kobayashi, T. Kurata, and S. Kato. Reactivities of some carbonyl compounds in Strecker degradation., *Agr. Biol. Chem.*, 32, 1968, 46-50.
27. J. Kerler, J.G.M. van der Ven, and H. Weenen. α-Acetyl-*N*-heterocycles in the Maillard reaction, *Food Rev. Int.*, 13 (1997) 553-575.
28. H. Weenen and S. B. Tjan. Analysis, structure, and reactivity of 3-deoxyglucosone, in: *Flavor precursors, thermal and enzymatic conversions*, R. Teranishi, G.R. Takeoka, and M. Güntert, Eds., ACS symposium series 490, ACS, Washington, D.C., 1992, 217-231.
29. H. Weenen and S.B. Tjan. 3-Deoxyglucosone as flavour precursor, in: *Trends in flavour research*, H. Maarse and D.G. van der Heij, Eds., Elsevier science B.V., Amsterdam, the Netherlands, 1994, 327-337.
30. A.F. Ghiron, B. Quack, T.P. Mawhinney, and M.S. Feather. Studies on the role of 3-deoxy-D-erythro-glucosulose (3-deoxyglucosone) in nonenzymatic browning. Evidence for involvement in a Strecker degradation, *J. Agric. Food Chem.*, 36, 1988, 677-680.
31. H. Weenen, J.G.M. van der Ven, L.M. van der Linde, J. van Duynhoven, and A. Groenewegen. C4, C5, and C6 3-deoxyosones: structures and reactivity, in: *The Maillard reaction in foods and medicine*, J. O'Brien, H.E. Nursten, M.J.C. Crabbe, and J. Ames, Eds., the Royal Society of Chemistry, Cambridge, U.K., 1998, 57-64.
32. H. Weenen. Reactive intermediates and carbohydrate fragmentation in Maillard chemistry, *Food Chem.*, 62 (1998) 393-401.

Chapter 16

Analysis of Low Molecular Weight Aldehydes Formed during the Maillard Reaction

John Didzbalis[1,2] and Chi-Tang Ho[2]

[1]Bestfoods Technical Center, 150 Pierce Street, Somerset, NJ 08873
[2]Department of Food Science, Rutgers University, 65 Dudley Road, New Brunswick, NJ 08901–8520

The detection of short chain aldehydes and other carbonyl compounds formed during the Maillard reaction poses a difficult challenge due to their high reactivity and volatility. Cysteamine (2-aminoethanethiol) was used to derivatize these carbonyl containing compounds in a model system of glucose/lysine to form stable thiazolidine compounds. The thiazolidine derivatives along with the other heterocyclic flavor compounds generated were isolated by high vacuum distillation and identified by GC/MS.

During the course of the Maillard reaction, a large pool of reactive short chain (C_1-C_4) carbonyl compounds are generated. These compounds serve as precursors for important flavor compounds, including pyrazines, thiazoles and other heterocyclic compounds. Retroaldolisation or β–scission of deoxyglycosones during both non-enzymatic browning and caramelization generate these compounds. They are also formed directly from carbohydrates via retroaldolisation. These fragments are very reactive and have been shown to possess much higher browning activity than a monosaccharide (*1*).

The combination of reactivity and volatility of these compounds make their analysis difficult. They would not be detected using traditional solvent extraction and concentration techniques. Since these carbonyl compounds serve as precursors of pyrazines, understanding their role throughout the course of the Maillard reaction would be of great interest.

The reaction of cysteamine (2-aminoethanethiol) and aldehydes to generate thiazolidine derivatives has been known for quite some time, Figure 1. The reaction was patented in 1975 to generate flavors for use in foods, beverages and tobacco products. The compounds were described to contribute meaty, sulfury, cereal, popcorn-like or nutty notes (*2*).

A more recent use for cysteamine has been as a derivatizing agent to allow for the detection of low molecular weight aldehydes. The method was based on the reaction of volatile carbonyl compounds with cysteamine to generate stable thiazolidine derivatives. The reaction could be performed under mild conditions, 25°C and neutral pH values, thus eliminating the need for cold trapping. The method has been used to determine formaldehyde in air and food samples (*3*). Subsequent research has examined the volatile carbonyl content in the headspace of both heated pork (*4*) and heated food oils (*5*).

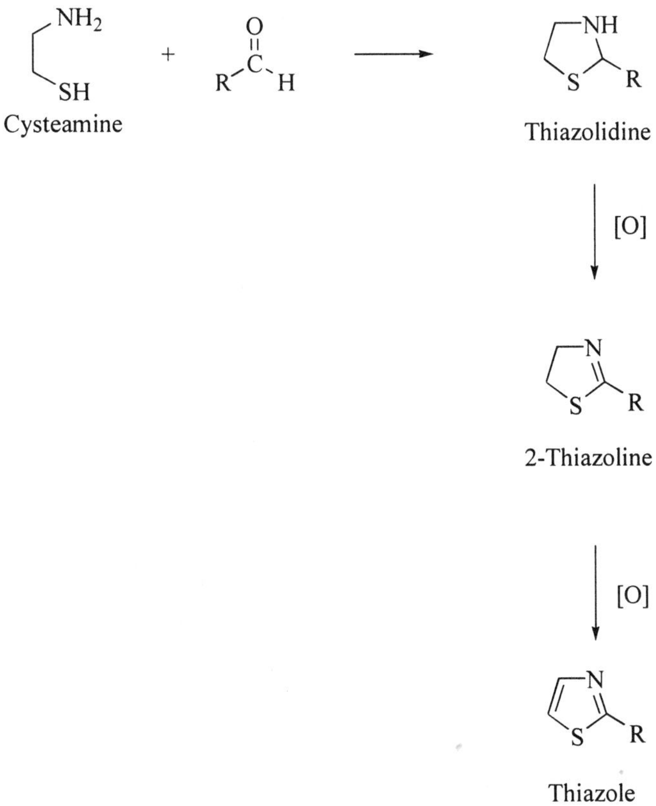

NH_2
SH
Cysteamine
+
O
R C H
NH
S R
Thiazolidine
[O]
N
S R
2-Thiazoline
[O]
N
S R
Thiazole

Figure 1. The reaction of carbonyl compound with cysteamine

The acetaldehyde content of foods and beverages (*6*), the carbonyl content of cigarette smoke (*7*) and automobile exhaust (*8*) was also measured using cysteamine derivatives. The technique has also been used to trap intermediates derived from heating carbohydrates and alanine (*9,10*).

Cysteamine can also be generated in foods by the thermal degradation of cysteine as shown in Figure 2, and can further react to form thiazoles, thiazolidines and thiazolines. These compounds have been found in heated model systems of glucose and cysteine (*11,12*).

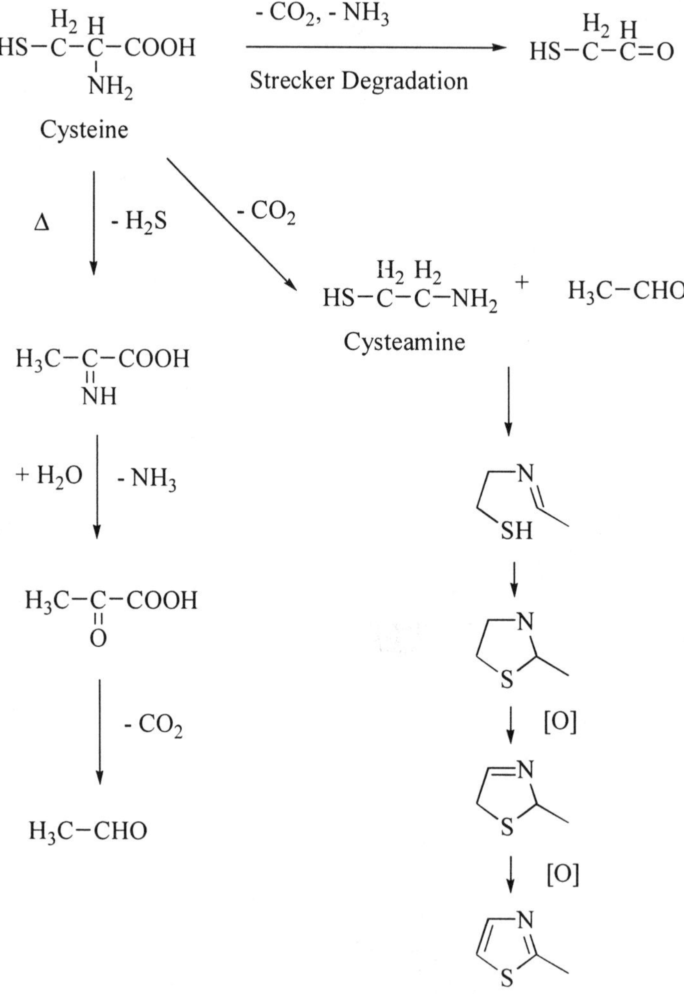

Figure 2. Generation of cysteamine from the decomposition of cysteine.

Recent attention has focused on acetylthiazolidines, which have been found to possess a pleasant popcorn like aroma (*13*). Huang and coworkers (*14,15*) have elucidated the mechanism for the formation of acetylthiazolidine and thiazines using model systems of cysteamine and diacetyl. At pH 8, cysteamine exists as a zwitterion.

The amino nitrogen (pK_a=10.7) of cysteamine attacks the carbonyl carbon of the carbonyl compound. Loss of the formed hydroxyl group gives a secondary carbocation. Another nucleophilic attack, this time by the thiolate on the carbocationic carbon leads to the substituted thiazolidine. Thiazole and thiazolines were postulated to form from the corresponding thiazolidine via dehydrogenation (*16*).

In this research, cysteamine was used as a derivatizing agent, to capture reactive carbonyl compounds generated during the heating of the glucose and lysine in an oil and water model system. This technique allowed for the analysis of not only the alkyl pyrazine products, but also gave insight into the composition of the pool of carbonyl containing pyrazine precursors.

Materials and Methods

L-lysine, D-glucose, 3-hydroxy-2-butanone, hydroxyacetone, 2,3-butanedione, pyruvaldehyde, glyceraldehyde, 2,3-pentanedione, 2,3-hexanedione, 2,5-dimethyl-4-hydroxy-3(2H)-furanone (furaneol), furfural, glyoxal and 4-methylpyrimidine were purchased from Aldrich Chemical Company (Milwaukee, WI). 2-aminoethanethiol was purchased from Fluka (Milwaukee, WI.). Diethyl ether without preservative and dichloromethane were purchased from Burdick and Jackson, Inc. (Muskegon, MI). Anhydrous sodium sulfate was acquired from Fisher Scientific (Fair Lawn, NJ). Soybean oil was purchased at a local supermarket.

Lysine/Glucose Cysteamine Models

To a 500 mL, 3-neck round bottom flask equipped with a water-cooled condenser, 3000 mg of L-lysine, 3690 mg of glucose and 150 mL ultra high purity water were added. Once the lysine and glucose had dissolved, 150 g of soybean oil was added. The reaction mixture was stirred vigorously using a mechanical stirrer while heated at 100°C for 1.5 hrs. Once the reaction had completed, the round bottom flask and attached condenser was cooled in an ice bath.

The total reaction mixture was divided into three equal fractions and stored at -60°C. The fractions were thawed and added to a 500 mL, 3-neck round bottom flask equipped with a water-cooled condenser. Cysteamine (1.58 g) was added to the round bottom, which was then stirred and heated at 60°C for 15 minutes. The round bottom was then cooled in an ice bath. The contents were then transferred to a 250-mL separatory funnel. The sample was extracted twice with 50-mL aliquots of dichloromethane. The dichloromethane phases were combined and stored at -60°C until the high vacuum distillation was performed.

Lysine/Glucose Heating Time Cysteamine Models

To measure the yield of both carbonyl compounds and pyrazines as a function of heating time, six models were prepared and reacted as follows. To the 500 mL, 3-neck round bottom flask equipped with a water-cooled condenser, 1.00 g of L-lysine, 1.23 g of glucose and 50 mL ultra high purity water were added. Once the lysine and glucose had dissolved, 50 g of soybean oil was added. The reaction mixture was stirred vigorously using a mechanical stirrer while heated at 100°C for the desired time period. The models were heated for 0.5, 1.0, 1.5, 2.0, 2.5 and 3.0 hrs.

Once the reaction had completed, the round bottom flask with the attached condenser was cooled in an ice bath. Cysteamine (1.58 g) was then added to the round bottom, which was stirred and heated at 60°C for 15 minutes. Upon completion, the round bottom was then cooled in an ice bath. The contents were then transferred to a 250-mL separatory funnel. The internal standard (4-methyl-pyrimidine) was added at this point. The sample was extracted twice with a 50-mL aliquot of dichloromethane. The dichloromethane phases were combined and stored at -60°C until the high vacuum distillation was performed.

Carbonyl Cysteamine Models

Known glucose degradation products were reacted with cysteamine to aid in the identification of the glucose/lysine generated cysteamine derivatives. In a 500 mL 3-neck round bottom flask equipped with a water-cooled condenser, 1000 mg lysine, the carbonyl compound, an equimolar amount of cysteamine and 50 mL ultra high purity H_2O were added. Once the compounds had dissolved, 50-mL soybean was added. The reaction was heated at 100°C for 45 minutes while stirred vigorously with a mechanical stirrer. The carbonyl compounds studied were 3-hydroxy-2-butanone, hydroxyacetone, glyceraldehyde, 2,3-pentanedione, 2,3-hexanedione, pyruvaldehyde, 2-furfuraldehyde, glyoxal, furaneol and diacetyl.

The high vacuum distillation technique developed by Schieberle and Grosch (*17*) was used to isolate the generated volatile flavor compounds in the model reactions. The technique was developed to isolate odor active compounds from foods and has been used to identify the aroma impact compounds in vegetable oils, butter, cheese, beef, bread, beer, tea, grapes, honey and trout. Off-flavors in lemon oil and soybean oil have also been isolated using the technique (*18,19*).

The technique is well suited for the analysis of flavor compounds from high fat systems. A high fat food is extracted with an organic solvent. The solvent, extracted fat and flavor compounds are then distilled under high vacuum, removing the flavor compounds and solvent and leaving the lipid material behind. The solvent can then be distilled, concentrating the flavor compounds.

The high vacuum apparatus used in this work consisted of a 500 mL, 2-neck round bottom flask to which a 250 mL separatory funnel with a ground glass joint was attached. A heated water jacketed distillation adapter attached to the round bottom (40°C) was also attached to a series of cold traps, immersed in Dewer flasks containing liquid nitrogen, -196°C. The cold traps were connected to a Toss 50 Pumping System, which consisted of a rotary vane vacuum pump and a turbomolecular pump (Leybold Vacuum Products, Export, PA).

The dichloromethane extract was added from the separatory funnel to the round bottom over the course of 1 hr, with the temperature of the round bottom and adapter held at 40°C. The oil was distilled under vacuum (10^{-6} mbar) at 40°C for an additional 3 hours.

Upon completion of the distillation, the ether collected in the two cold traps was thawed, combined and dried over anhydrous sodium sulfate. The ether extract was filtered using 2V filter paper (Whatman #1202 150, Fisher Scientific) into a 100 mL pear shaped flask. The extract was concentrated to 5 mL using a vacuum jacketed Vigreux column (15 mm x 500 mm) in water bath heated to 40°C. The extract was transferred to a 2 mL pear shaped flask and further concentrated to 500 µL using a micro distillation apparatus.

Compound identification and quantitation was done using a Finnigan 4600 Quadrupole Mass Spectrometer interfaced to a Varian 3400 GC. The GC conditions for the DB5 column (60 m x 0.25 mm id x 0.25 µm film thickness) were 35°C held for 2 min, heated at 30°C/min to 60°C and held for 1 min, followed by heating at 6°C/min to 250°C and held for 10 min. The sample (0.5 µL) was injected using an on-column injection technique with the injector programmed to track the oven temperature.

The mass spectrometer electron ionization (EI) was set at 70 eV. The ion source temperature was 150°C, with the emission current set at 0.3 mA, scanning masses 35-400. Mass spectra was identified by utilizing on-line computer libraries, Wiley Registry of Mass Spectral Data 5th Ed. and the NIST/EPA/NIH Mass Spectral Library (1996), and published mass spectra.

Chemical ionization (CI) using CH_4 as the reagent gas was performed to gain additional structural and molecular weight information where EI spectra alone was not sufficient for compound identification. Linear retention indices for the compounds were calculated using straight chain alkanes (Hydrocarbons C_7-C_{30}, Aldrich), using the method of Majlat et al. (*20*). The volatile flavor compounds were quantified based on the area response of the added internal standard.

Results and Discussion

To test the potential of this method for analysis of carbonyl products, a model of glucose/lysine was reacted for 1.5 hours at 100°C and divided into thirds. The samples were stored frozen at –60°C and reacted individually. The results seen in Table 1, show that the technique was successful in derivatizing the carbonyl fragments. Along with the expected alkylpyrazines, unsubstituted pyrazine, 2-methylpyrazine, 2,5-dimethylpyrazine, 2,3-dimethylpyrazine, 2-ethyl-6-methylpyrazine and trimethyl-pyrazine, a series of thiazolidine derivatives were detected. An additional heterocyclic compound, trimethyloxazole was also detected. Trimethyloxazole can be obtained from aldehydes and α-aminoketones via oxazolines (*21*). It can also be generated from acetoin in the presence of NH_3 .

Thiazoline, 2-methylthiazoline, thiazolidine, 2-methylthiazolidine, 2,2-dimethylthiazolidine, 2-ethylthiazolidine, 2-ethyl-2-methylthiazolidine, 2-acetyl-thiazoline, 2-methyl-2-hydroxymethylthiazolidine, 2-acetyl-2-methylthiazolidine and 2-pentylthiazolidine were detected. The thiazolidine derivatives were identified using published EI data (*22*) and library data. CI spectra were also utilized for molecular weight determination when required.

The consistent pyrazine concentration values indicated that storing the reacted models at –60°C did not cause much degradation of the sample. The concentration of thiazoline and thiazolidine are shown with the parent carbonyl compound, formaldehyde. The concentration of 2-ethyl-2-methylthiazolidine corresponds to methyl ethyl ketone and 2-ethylthiazolidine corresponds to propanal. The concentration of 2-acetyl-2-methylthiazolidine is shown as diacetyl. Pyruvaldehyde corresponds to 2-acetyl-2-thiazoline. Again, 2-methylthiazoline and 2-methylthiazolidine correspond to the parent carbonyl compound, acetaldehyde. Acetone and hydroxyacetone correspond to 2,2-dimethylthiazolidine and 2-methyl-2-hydroxymethylthiazolidine.

Carbonyl Cysteamine Models

As discussed earlier in the Materials Section, a series of carbonyl containing compounds were reacted with cysteamine to generate the thiazolidine derivative. Lysine was added to the model to adjust the pH and make the model as close as possible to the glucose/lysine models. This was done to examine the effectiveness of the derivatization reaction under the experimental conditions and also to measure and identify the major products of the derivatization of α-dicarbonyl and other carbonyl containing fragments. Research has shown that one can generate different thiazolidine derivatives in a single reaction. As an example, both thiazolidine and thiazines could be formed from the same cysteamine/diacetyl model system by varying the reaction conditions (*23*).

In the (3-hydroxy-2-butanone) acetoin/cysteamine model, the only product generated was 2-methyl-2-acetylthiazolidine. The hydroxyacetone/cysteamine model generated additional thiazolidine derivatives, than the expected 2-methyl-2-hydroxymethylthiazolidine. The compound in highest yield was 2-methyl-thiazolidine. Smaller amounts of 2-methyl-2-hydroxymethylthiazolidine, 2-ethylthiazolidine, 2-methylthiazoline, 2-acetylthiazolidine, 2-acetylthiazoline and 2-methylpyrazine were also detected.

In the glyceraldehyde/cysteamine model, 2-acetyl-2-thiazoline was generated in the highest yield. Again small amounts of 2-methylthiazoline, thiazolidine, 2-methylthiazolidine, 2-ethylthiazolidine, 2-propylthiazolidine, 2-acetylthiazolidine, 2-acetylthiazole, 2,5-dimethylpyrazine and trimethylpyrazine were generated.

In the 2,3-pentanedione/cysteamine model, the major products generated were ethyl-2-(2-methylthiazolidine) ketone and 2-ethyl-2-acetylthiazolidine. Small amounts of 2-methylthiazolidine and 2-ethylthiazolidine were detected. Similar results were seen in the 2,3-hexanedione sample, which generated *n*-propyl-2-(2-methylthiazolidine) ketone and 2-propyl-2-acetylthiazolidine with similar minimal amounts of 2-methylthiazolidine and 2-ethylthiazolidine.

Table I. Reproducibility of Derivatizing the Reaction Mixture of Glucose/Lysine with Cysteamine

		Time at -60°C (Days)	*0*	*5*	*10*
RI	*Compound*	*Parent Carbonyl*	*Conc. (µg/g lysine)*		
733	Pyrazine		321.5	279.1	277.5
819	Thiazoline	Formaldehyde	15.0	11.6	13.0
824	Methylpyrazine		1246.1	1119.9	1122.2
880	2-methylthiazoline	acetaldehyde	130.7	140.8	90.0
902	Thiazolidine	formaldehyde	72.5	146.1	115.4
905	2,5-dimethylpyrazine		981.0	1289.0	1002.0
911	2,3-dimethylpyrazine		124.0	123.0	120.0
932	2-methylthiazolidine	acetaldehyde	8924.8	9668.1	10265.7
952	2,2-dimethylthiazolidine	acetone	943.0	684.0	1164.2
1010	2-ethyl-6-methylpyrazine		0.7	0.7	1.0
1013	Trimethylpyrazine		52.1	57.8	65.7
1024	2-ethylthiazolidine	propanal	3.8	3.0	3.5
1058	2-ethyl-2-methylthiazolidine	methyl ethyl ketone	11.8	9.1	16.0
1109	2-acetyl-2-thiazoline	pyruvaldehyde	22.6	78.9	19.1
1126	2-methyl-2-hydroxymethyl-Thiazolidine	hydroxyacetone	1238.4	568.2	328.6
1155	2-acetyl-2-methyl-thiazolidine	diacetyl	85.9	108.2	87.8
1344	2-pentylthiazolidine	hexanal	6.4	5.4	7.1
1410	3,4-dithiahexyl-1,6-diamine	cysteamine dimer	2802.1	2501.7	1367.7

The pyruvaldehyde/cysteamine model, generated 2-acetyl-2-thiazoline in the highest yield. Additional components detected in trace levels included, 2-acetylthiazolidine, 2-acetylthiazole, 2-methyl-2-thiazoline, thiazolidine, 2,5-dimethylpyrazine and trimethylpyrazine. The glyoxal (ethanedial)/cysteamine model, yielded equal amounts of 2-methylthiazoline, 2-methylthiazolidine and 2-hydroxymethyl-2-thiazoline.

The 2,3-butanedione (diacetyl)/cysteamine model yielded only 2-methyl-2-acetylthiazolidine. The 2-furfural/cysteamine model generated 2-fufurylthiazolidine and the furaneol/cysteamine model yielded no products.

The small concentration of generated pyrazines in some models showed the high degree of reactivity of carbonyl fragments with lysine. The fragments reacted with lysine at a faster rate than the cysteamine reaction to form the thiazolidine derivative.

Additional models found that aldehydes derived from lipid oxidation reacted with cysteamine to form the corresponding thiazolidine derivatives. Pentanal, hexanal, heptanal, 2-heptenal, 2,4-heptadienal, octanal, nonanal and 2,4-decadienal generated 2-butylthiazolidine, 2-pentylthiazolidine, 2-hexylthiazolidine, 2-(1,3-hexadiene)-thiazolidine, 2-heptylthiazolidine, 2-octylthiazolidine and 2-(1,3-nonadiene)-thiazolidine.

Since clean soybean oil was used in the cysteamine derivatization of the glucose/lysine models, only the hexanal derivative, 2-pentylthiazolidine was detected.

Glucose/Lysine Heating Time Cysteamine Study

The results of cysteamine derivatization of carbonyl compounds generated over the course of heating glucose/lysine at 100°C for 3 hours are seen in Table II. Individual models were prepared for each point and heated for the required 30-minute interval. The generated alkyl pyrazines and thiazolidine compound, along with the parent carbonyl are shown in the table.

The alkyl pyrazines detected included unsubstituted pyrazine, methylpyrazine, 2,5-dimethylpyrazine, 2,3-dimethylpyrazine, 2-ethyl-6-methylpyrazine and trimethyl-pyrazine.

The concentration of methylpyrazine and 2,5-dimethylpyrazine increased as a function of heating time until a plateau was reached at 2.5 hrs. In the early heating stages, the concentration of both pyrazines was comparable with the concentration of methylpyrazine slightly higher at the plateau region. A similar trend was seen for 2,3-dimethylpyrazine and 2-ethyl-6-methylpyrazine. The yield of methylpyrazine, 2,5-dimethylpyrazine, 2,3-dimethylpyrazine and 2-ethyl-6-methylpyrazine appear to follow a similar shaped curve, an initial increase, followed by a plateau region. In similar work, Jusino and coworkers (*24*) examined the reaction order for pyrazine formation in model systems of amioca starch, lysine and glucose. They used a fractional conversion technique to measure the formation kinetics of 2,5-dimethyl-pyrazine and methylpyrazine. They concluded that given sufficient time the concentrations of 2,5-dimethylpyrazine and methylpyrazine reached a plateau. They found that the reaction followed a first order reaction. They also concluded that methylpyrazine and 2,5-dimethylpyrazine were governed by a similar rate determining step.

Pyrazine with a lower E_{act} appeared to still be increasing and had not yet reached a plateau as was seen for the other pyrazine compounds. Trimethylpyrazine appeared to deviate from the expected trend reaching a maximum concentration and then showing a decline.

The parent carbonyl compounds detected included formaldehyde, acetaldehyde, acetone, propanal, methyl ethyl ketone, pyruvaldehyde, hydroxyacetone and diacetyl.

The data for formaldehyde and diacetyl show that diacetyl was generated at a faster rate than formaldehyde. Diacetyl has a sinusoidal type curve with maxima at 1 and 2.5 hours. Formaldehyde shows an increase with a maximum at 2.5 hrs followed by a reduction in concentration. Propanal and methyl ethyl ketone show more typical behavior with a gradual increase in concentration followed by a more constant plateau. The concentration of these two carbonyls is much lower than the concentrations of formaldehyde and diacetyl.

Table II. Compounds Generated from Derivatization of Carbonyl Compounds Over the Course of Heating Glucose/Lysine

Heating time glu + lys (hr)	*0.5*	*1.0*	*1.5*	*2.0*	*2.5*	*3.0*
Compound			*Concentration (μg/g lysine)*			
Pyrazine	95.61	125.95	292.7	259.28	384.6	572.7
Thiazoline	11.2	10.0	15.0	10.0	0.0	12.0
methylpyrazine	393.2	800.5	1246.1	1420.5	1332.4	1246.0
trimethyloxazole	nd	nd	nd	58.3	24.3	26.2
2-methylthiazoline	137.4	85.2	130.7	94.0	85.0	66.8
thiazolidine	43.7	29.6	72.5	55.3	155.9	46.0
2,5-dimethylpyrazine	502.0	870.9	981.0	1333.3	849.5	839.0
2,3-dimethylpyrazine	62.3	55.0	124.0	88.2	90.0	92.0
2-methylthiazolidine	9958.6	10583.0	8924.8	11921.2	8737.5	9402.8
2,2-dimethylthiazolidine	377.0	959.2	943.0	1135.7	1002.3	695.5
2-ethyl-6-methylpyrazine	nd	nd	0.7	1.2	2.1	1.8
trimethylpyrazine	10.7	54.6	52.1	125.1	90.9	42.8
2-ethylthiazolidine	8.6	14.5	3.8	18.2	11.4	8.8
2-ethyl-2-methylthiazolidine	2.6	8.5	11.8	11.5	15.6	13.1
2-acetyl-2-thiazoline	7.4	20.9	22.6	43.6	35.2	17.3
2-methyl-2-hydroxymethyl-thiazoline	534.0	1789.7	1238.4	3158.5	1901.6	653.4
2-acetyl-2-methyl-thiazolidine	104.7	240.9	85.9	199.6	250.9	122.7
2-pentyl thiazolidine	5.3	5.6	6.4	2.2	1.0	6.0
3,4-dithiahexyl-1,6-diamine	2124.8	1411.1	2802.1	1162.9	1566.8	1985.0

The highest concentration of the carbonyl compounds detected was acetaldehyde, followed by hydroxyacetone and acetone. Acetaldehyde showed a rapid increase in concentration followed by a relativelyconstant steady state. The absence of any other 2 carbon thiazolidine derivatives as seen in the glyoxal model, would indicate that the primary source of 2-methylthiazolidine was acetaldehyde. Hydroxyacetone showed a maximum at 2.5 hours followed by a decline. Acetone showed a similar trend, also achieving a maximum at 2.5 hours.

The results showed the feasibility of using cysteamine to derivatize carbonyl containing Maillard precursors in the glucose/lysine model. The volatility and reactivity of these carbonyl compounds make them then extremely difficult to analyze. The results demonstrated that cysteamine could be used as a trap to derivatize the carbonyl containing intermediates, which could then be analyzed in a traditional analytical technique along with the resulting flavor compounds. Further work is required to optimize the derivatization technique to ideally yield a single thiazolidine derivative from a single reactant. The cysteamine-carbohydrate fragment models showed that in some cases more than one thiazolidine derivative was formed from a single reactant.

References

1. Hayashi, T.; Namiki, M. *Agric Biol Chem.* **1986**, 50, 1965-1970.
2. Flament, I. US Patent 3,881,025, 1975.
3. Shibamoto, T. In *Flavors and Off-Flavors '89;* Charalambous, G., Ed.; Developments in Food Science; Elsevier Science B.V.: New York, 1990; Vol. 24, pp 471-484.
4. Yasuhara, A.; Shibamoto, T. *J. Food Sci.* **1989**, 54, 1471-1472, 1484.
5. Yasuhara, A.; Shibamoto, T. *J. Chromatogr.* **1991**, 547, 281-298.
6. Miyake, T.; Shibamoto, T. *J. Agric. Food Chem.* **1993**, 41, 1968-1970
7. Miyake, T.; Shibamoto, T. *J. Chromatogr.* **1995**, 693, 376-381.
8. Yasuhara, A.; Shibamoto, T. *J. Chromatogr.* **1994**, 672, 261-266.
9. Weenan, H. *Food Chemistry.* **1998**, 62, 393-401.
10. Weenan, H.; Tjan, S.B.; de Valois, P.J.; Vonk, H. In *Thermally Generated Flavors*; Parliament, T.H.; Morello, M.J.; McGorrin, R.J., Eds.; ACS Symp. Ser. 543; American Chemical Society: Washington, DC, 1994, 142-157.
11. Scanlon, R.A.; Kayser, S.G.; Libbey, L.M.; Morgan, M.E. *J. Agric. Food Chem.* **1973**, 21, 673-675.
12. Mulders, E.J. *Z. Lebensm. Unters. Forsch.* **1973**, 152, 193-201.
13. Umano, K.; Hagi, Y.; Nakahara, K.; Shyoji, A.; Shibamoto, T. *J. Agric. Food Chem.* **1995**, 43, 2212-2218.
14. Huang, T.C.; Huang, L.Z.; Ho, C.-T. *J. Agric. Food Chem.* **1998**, 46, 224-227.
15. Huang, T.C.; Su, Y.M.; Ho, C.-T. *J. Agric Food Chem.* **1998**, 46, 664-667.
16. Sheldon, S.A.; Shibamoto, T. *Agric. Biol. Chem.* **1987**, 51, 2473-2477.
17. Schieberle, P.; Grosch, W. *Fette Seifen Anstrichm.* **1985**, 87, 76-80.
18. Grosch, W. *Flavor and Fragrance Journal.* **1994**, 9, 147-158.
19. Schieberle, P. In *Characterization of Food: Emerging Methods;* Gaonkar, A.G., Ed.; Elsevier Science B.V., New York, 1995.
20. Majlat, P.; Erdos, Z.; Takacs, J. *J. Chromo.* **1974**, 91, 89-103.

21. Vernin, G.; Parkanyi, C. In *Heterocyclic Flavouring and Aroma Compounds*; Vernin, G., Ed.; Ellis Horwood Publishers: Chicester, U.K., 1982; pp151-201.
22. Yasuhara, A.; Kawada, K.; Shibamoto, T. *J. Agric. Food Chem.* **1998**, 46, 2664-2670.
23. Huang, T.C.; Su, Y.M.; Ho, C.-T. *J. Agric Food Chem.* **1998**, 46, 664-667.
24. Jusino, M.G.; Ho, C.-T.; Tong, C.H. *Agric Food Chem.* **1997**, 45, 3164-3170.

Chapter 17

Chiral 1,3-Octanediol Synthetic Studies Using Enzymatic Methods as Key Steps

M. Nozaki, M. Ikeuchi, and N. Suzuki

Central Research Laboratory, Takasago International Corporation, 1–4–11, Nishi-Yawata, Hiratuka, 254–0073, Japan

Synthesis of both enantiomers of 1,3-octanediol was accomplished via chiral 1-octen-3-ol. Enzymatic optical resolution was adopted as the key step for obtaining chiral 1-octen-3-ol. Both enantiomers of 1,3-octanediol were transformed into their glucosides enzymatically. The acetals of the diol were also synthesized. Enzymatic reaction played a major role in the synthesis of chiral 1,3-octanediol and its derivatives. Sensory evaluation was performed on all of the samples.

It has been reported that (*R*)-1,3-octanediol β-glucoside is a major constituent of apple glycosides (*1*). It is reported that these glycosides are aroma precursors of apple flavors. It has also been established that the acetaldehyde acetal of the diol is responsible for a key green note of apple aroma (*2*). Therefore it can be assumed that (*R*)-1,3-octanediol plays a major role in apple flavor formation. The aim of this study was to clarify the sensory properties of 1,3-octanediol stereoisomers and their derivatives. The stereoisomers included enantiomers, anomers and diastereomers. To clarify sensory properties, adequate amounts of samples with strictly defined stereochemistry were needed. To accomplish this aim, it was required that the synthesis is easy to handle and the products have strictly defined stereochemistry. It was thought that enzymatic reactions could satisfy these requirements. The synthesis where enzymatic reactions are key steps is called hybrid synthesis. In this study, hybrid synthesis was adopted.

Synthesis of racemic 1,3-octanediol has been reported. It was also reported that synthesis of *R*- and *S*-form of the diol was accomplished by use of Sharpless oxidation of (*E*)-2-octenol as a key step. The efficiency of the synthesis was very low and sensory properties of the enantiomers have not been reported. Thus we intended to synthesize optically pure enantiomers and their derivatives conveniently and to perform sensory evaluation of these. Convenience of the synthesis was most important.

We have tried optical resolution of aroma active secondary alcohols by the use of lipases and succeeded in the enzymatic optical resolution of 1-octen-3-ol (*3*). We decided to use optically active 1-octen-3-ol as a key intermediate to synthesize optically active 1,3-octanediol.

Both enantiomers of 1-octen-3-ol were chemically transformed into the corresponding enantiomers of the diol. Products were subjected to sensory evaluation.

Glucosidation of racemic 1,3-octanediol by Königs-Knorr reaction has been previously reported (*1*). However, the results were not satisfactory in yield and anomeric specificity. Regarding some glucosides, for instance, ethyl glucoside, it has been known that there is difference in taste among anomers. To evaluate both anomers of 1,3-octanediol glucoside, anomer specific glucosidation was needed. Enzymatic glucosidation was examined. Four glucosides were synthesized enzymatically. Those were the α– and β-glucosides of *R*- and *S*-diol. There was a significant difference in taste among those four isomers.

The acetals of the diol enantiomers (2-methyl-4-pentyl-1,3-dioxane) were also synthesized. This acetal was found in aged apple cider and provides the key green note of apple cider. The absolute configuration of the diol is 3*R* and *cis* form dioxane exists as a major component. The acetal has two chiral centers and all four diastereomers were synthesized. Sensory evaluation of the diastereomers revealed that there were significant differences in aroma between them.

Synthesis of optically active 1,3-octanediol

1-octen-3-ol was chosen as the starting material. Because (*R*)-1-octen-3-ol is a very useful flavoring substance, asymmetric and hybrid synthesis was attempted. It was very difficult to obtain optically pure 1-octen-3-ol by asymmetric synthesis, but enzymatic optical resolution of the racemates by use of lipases gave excellent results. The optically pure enantiomers obtained were transformed into the corresponding chiral 1,3-octanediol.

It has been reported that 1-octen-3-ol, the character impact compound of mushrooms, occurs predominantly in the *R* configuration in mushrooms and that there are different sensory properties between enantiomers. Optical resolution of racemic 1-octen-3-ol by PPL (porcine pancreatic lipase) was previously reported. However, the given (*R*)-1-octen-3-ol had 60% *ee*. Other lipases were tested to try to improve the enantiomeric resolution. Among lipases examined, CAL (*Candida antarctica* lipase) gave the best result. Two grams of racemic 1-octen-3-ol and 6.7 grams of vinyl acetate were added to 1 gram of CAL in 100 ml of hexane. The reaction mixture was stirred at 25°C for 3 hours and the conversion attained was 51%. (*S*)-1-octen-3-yl acetate was afforded at the yield of 90% and 95% *ee* after chromatographic separation on silica gel. The unreacted *R*-alcohol remaining in the reaction mixture was given at the yield of 89% and 98% *ee* after the silica gel chromatographic isolation (Fig. 1) (*3*).

Enzymatic optical resolution of 1-octen-3-ol

(*R,S*)-1-octen-3-ol → [Lipase, vinyl acetate (OAc), n-hexane] → (*S*)-1-octen-3-yl acetate + (*R*)-1-octen-3-ol

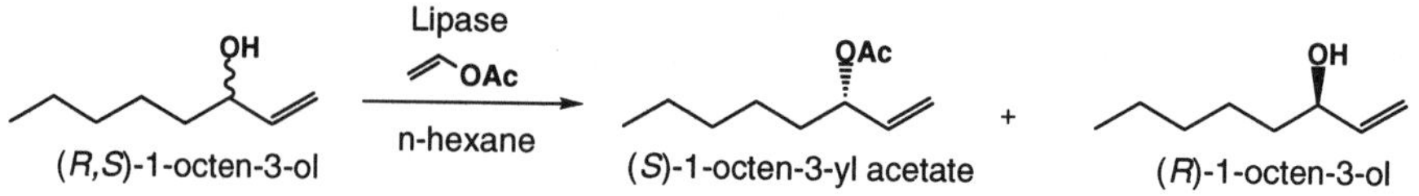

lipase	*conv. % (h)*	*(R)-form ee (%)*	*E value*[a]
Pseudomanas sp. lipase	53 (6)	33	2.5
Pseudomonas cepacia lipase	50 (24)	54	5.2
Alcaligenes sp. lipase	54 (24)	36	2.6
Candida antarctica lipase	51 (18)	96	97.3

[a] E value: enantiomeric ratio[4)]

Figure 1. Enzymatic optical resolution of 1-octen-3-ol.

Table I. Sensory properties of chiral 1-octen-3-ol.

compound	*odor description*	*threshold in water*
(*R*)-1-octen-3-ol	intensive mushroom, fruity, green	1 ppm
(*S*)-1-octen-3-ol	herbaceous, musty, weak mushroom	10 ppm

Each enantiomer was further purified by recrystallization of its dinitrobenzoyl ester (99.5% *ee*). Both purified enantiomers were sensorily evaluated (Table I).

Transformation of chiral 1-octen-3-ol into chiral 1,3-octanediol

(*R*)-1-octen-3-ol was treated with *t*-butyldimethylsilylchloride and imidazole in DMF and gave the protected alcohol at 96% yield. The given TBS ether was treated with BF_3 in THF and then $NaOH/H_2O_2$(hydroboration/oxidation) and gave (3*R*)-*t*-butyldimethylsililoxyoctanol at 74% yield. This was deprotected with HF in CH_3CN and gave (*R*)-1,3-octanediol at 75% yield. The *ee* of the obtained diol was 99.5% (Fig. 2). By the same way, (*S*)-1-octen-3-ol gave (*S*)-1,3-octanediol. The purity of the enantiomers was measured by HRGC as MTPA (α-methoxy-α-(trifluoromethyl) phenylacetic acid) esters. The optical rotation of *R*-diol was $[\alpha]^{25}_D = +1.2(c=1.15, CHCl_3)$. *S*-diol was $[\alpha]^{25}_D = -1.2(c=1.00, CHCl_3)$. At the same time asymmetric synthesis was examined. Ethyl 3-oxooctanoate was hydrogenated with BINAP (2,2'-bis(diphenylphosphino)-1,1'-binaphthyl)-metal catalyst and gave ethyl (*R*)-3-hydroxyoctanoate. This was treated with lithium aluminum hydride and gave (*R*)-1,3-octanediol. Each enantiomer was further purified by recrystallization of its dinitrobenzoyl ester (99.5% *ee*). Sensory evaluation was performed on both purified enantiomers (Table II).

Table II. Sensory properties of chiral 1,3-octanediol.

compound	*odor description*
(*R*)-1,3-octanediol	sweet, astringent, body
(*S*)-1,3-octanediol	bitter, astringent

Enzymatic glucosidation of optically active 1,3-octanediol

It has been known that enzymatic glucosidation proceeds with anomeric specificity and that protection and deprotection steps are not required. Convenience of the reaction was one of the most important considerations. Enzymatic glucosidation was considered to be most suitable. Table III shows the results of using various commercially available enzymes for β-transglucosidation.

Many combinations of enzymes and glucose donors were examined. The combination of Cellulosin T2 and cellobiose as glucose donor gave the best results. Table IV shows the optimized conditions for β-glucosidation.

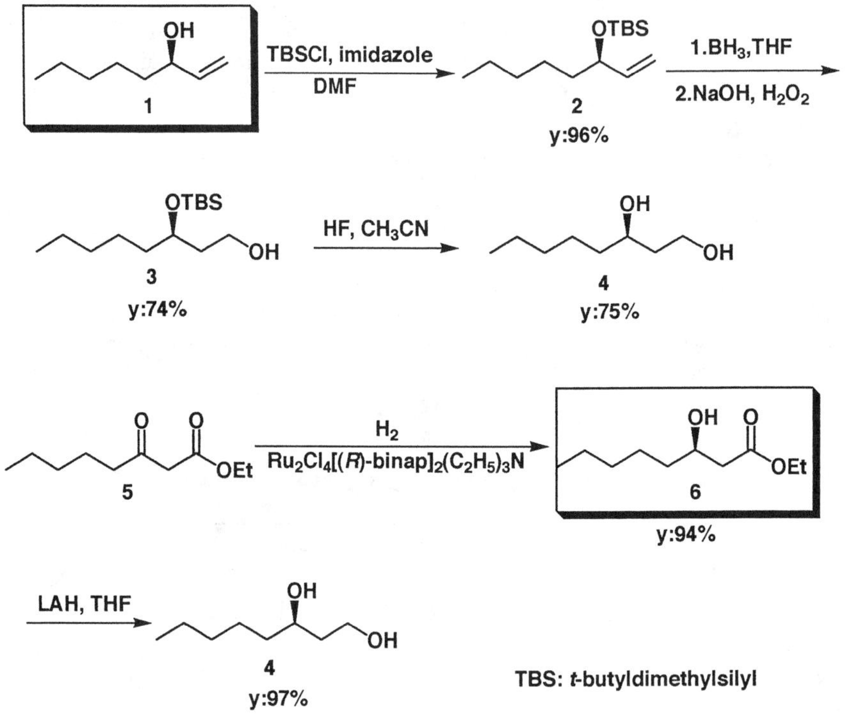

OH

TBSCl, imidazole

DMF

1

OTBS

2

y:96%

1.BH_3,THF

2.NaOH, H_2O_2

OTBS

OH

3

y:74%

HF, CH_3CN

OH

OH

4

y:75%

O O

OEt

5

H_2

$Ru_2Cl_4[(R)\text{-binap}]_2(C_2H_5)_3N$

OH O

OEt

6

y:94%

LAH, THF

OH

OH

4

y:97%

TBS: *t*-butyldimethylsilyl

Figure 2. Synthesis of (R)-1,3-octanediol.

Table III. Relative β-transglucosidase activity of various enzymes.

Enzyme	*Relative activity*
Meicelase	100
Cellulase AMANO A	54
Cellulase AMANO T	82
ONOZUKA R10	106
Cellulosin T2	112
Emulsin	38
NOVOZYM	40
Sumizyme C	54

Table IV. Optimized conditions for β-glucosidation.

Enzyme	Cellulosin T2 (1.2 mg/ml)
Donor	cellobiose 5% (W/W)
Acceptor	1,3-octanediol 5% (W/W)
Buffer	0.05M acetate buffer (pH 5.3, 5% DMF)
Temp.	45°C
Reaction Time	48 hrs (stirred)

The reaction proceeded smoothly and was terminated by separation of the upper and lower layer of the reaction mixture. The upper layer was fractionated by silica gel column chromatography and gave the desired β-glucoside and unreacted alcohol. The isolated glucoside was recrystallized from ethanol. Sensory evaluation was performed on the purified β-glucosides. (*R*)-1-β-glucopyranosyl-3-octanol and (*S*)-1-β-glucopyranosyl-3-octanol were obtained. Under these conditions, 50 grams of aglycon gave 9.0 grams of β-glucoside. There was no difference in yield between *R*- and *S*-diol glucosidation.

Table V shows the results of using various commercially available enzymes for α-transglucosidation.

Many combinations of enzymes and glucose donors were examined. The combination of Transglucosidase AMANO and maltose or isomaltose gave the best results. Table VI shows the optimized conditions for α-glucosidation. Under these conditions, 50 grams of aglycon gave 4.1 grams of α-glucoside. There was no difference in yield between *R*- and *S*-diol glucosidation.

Table V. Relative α-transglucosidase activity of various enzymes.

Enzyme	*Relative activity*
Transglucosidase AMANO	100
Contizyme	35
CGTase	50
Biozyme	10

Table VI. Optimized conditions for α-glucosidation.

Enzyme	Transglucosidase AMANO (12 μl/ml)
Donor	maltose or isomaltose 5% (W/W)
Acceptor	1,3-octanediol
Buffer	0.05M acetate buffer (pH 5.3, 5% DMF)
Reaction Temp.	45°C
Reaction Time	24 hrs (stirred)

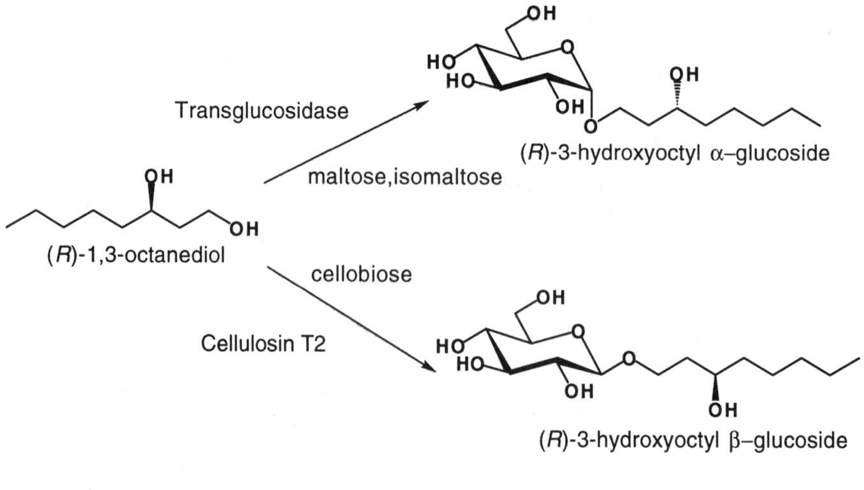

Figure 3. Anomer specific glucosidation of (R)-1,3-octanediol.

The reaction conditions were the same as for β-glucosidation. (*R*)-1-α-glucopyranosyl-3-octanol and (*S*)-1-α-glucopyranosyl-3-octanol were obtained. Sensory evaluation was performed on the purified α-glucosides. Fig. 3 shows anomer specific glucosidation of (*R*)-1,3-octanediol. Glucosidation at the 3-position of the diol was not observed.

The four glucosides were subjected to sensory evaluation. Table VII shows the results of sensory evaluation of the four glucosides at concentrations of 10 ppm in water.

Table VII. Sensory properties (taste description) of 1,3-octanediol glucosides.

glucosides	*sensory properties*
(*R*)-1,3-diol α-glucoside	sweet, warm, fresh
(*S*)-1,3-diol α-glucoside	sweet, bitter, astringent, warm
(*R*)-1,3-diol β-glucoside	sweet, bitter, warm
(*S*)-1,3-diol β-glucoside	sweet, astringent, warm

Both α-glucosides of *R*- and *S*-diol had better taste characteristics than the β-glucosides. The order of taste intensity was as follows: the strongest was α-glucoside of *S*-diol, followed by β-glucoside of *S*-diol, β-glucoside of *R*-diol and α-glucoside of *R*-diol. It was very interesting that non-naturally occurring diol glucosides had the more potent taste.

Transformation of optically active 1,3-octanediol to acetaldehyde acetal

(*R*)-1,3-octanediol was treated with acetaldehyde according to Schwab's method to give a diastereomeric mixture of the acetal. The diastereomers were easily separated by silica gel chromatography to yield pure diastereomeric dioxanes (2*S*,4*R* and 2*R*,4*R*). 2*S*,4*R*-dioxane is the predominant isomer occurring in nature. The same procedures were applied to (*S*)-1,3-octanediol, which does not occur in nature, to yield diastereomeric dioxanes (2*S*,4*S* and 2*R*,4*S*). Sensory evaluation was performed on these four dioxanes. Table VIII shows the sensory properties of these compounds.

There were significant differences in odor character, intensity and optical rotation among these four dioxanes. Naturally abundant dioxane and its mirror image had green character, while the less abundant dioxane and its mirror image had earthy character. In enantiomeric pair, naturally occurring dioxane had fresher character.

Conclusion

1,3-Octanediol and its derivatives were synthesized with strictly defined

Table VIII. Sensory properties and threshold values (in water) of the dioxanes.

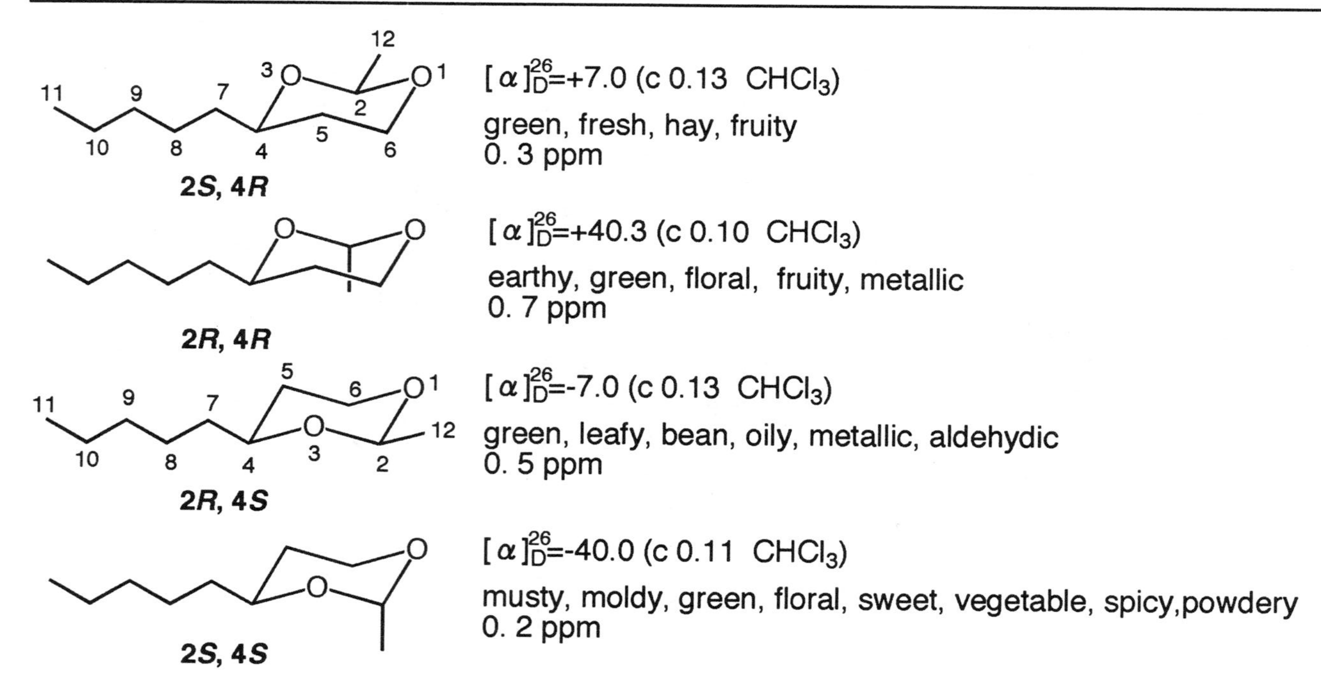

Isomer	Optical rotation	Odor	Threshold
2*S*, 4*R*	$[\alpha]_D^{26}$=+7.0 (c 0.13 $CHCl_3$)	green, fresh, hay, fruity	0. 3 ppm
2*R*, 4*R*	$[\alpha]_D^{26}$=+40.3 (c 0.10 $CHCl_3$)	earthy, green, floral, fruity, metallic	0. 7 ppm
2*R*, 4*S*	$[\alpha]_D^{26}$=-7.0 (c 0.13 $CHCl_3$)	green, leafy, bean, oily, metallic, aldehydic	0. 5 ppm
2*S*, 4*S*	$[\alpha]_D^{26}$=-40.0 (c 0.11 $CHCl_3$)	musty, moldy, green, floral, sweet, vegetable, spicy,powdery	0. 2 ppm

stereochemistry. Hybrid type synthesis was adopted. The key steps were enzymatic optical resolution of 1-octen-3-ol and enzymatic glucosidation of 1,3-octanediol. Adequate amounts of samples for sensory evaluation could be obtained conveniently. It was proven that this synthetic study contributed to measurement of sensory characteristics of aroma chemicals which were not easily obtained or their antipodes' character had not been known.

Acknowledgement

We thank Dr. Matsuda and Mr. Oshikubo for helpful discussions.

References

1. Schwab, W.; Schreier, P. *J. Agric. Food. Chem.* **1990**, *38,* 757-763.
2. Dietrich, C.; Schwab, W. *J. Agric. Food. Chem.* **1997**, *458,* 3178-3182.
3. Nozaki, M.; Suzuki, N.; Oshikubo, S. In *Flavour Science-Recent Developments,* eds. Taylor A.J. and Mottram D.S., The Royal Society of Chemistry, Cambridge, UK, **1996**, pp 168-171.
4. Chen, C.-S.; Fujimoto, Y.; Girdaukas, G., Sih, C. J. *J. Am. Chem. Soc.* **1982,** *104*, 7249-7299.

Chapter 18

Formation Pathways of 3-Hydroxy-4,5-dimethyl-2[5H]-furanone (Sotolon) in Citrus Soft Drinks

W. Schwab, T. König, B. Gutsche, M. Hartl, R. Hübscher, and P. Schreier

Lehrstuhl für Lebensmittelchemie, Universität Würzburg, Am Hubland, 97074 Würzburg, Germany

After storage of citrus soft drinks supplemented with a vitamin mixture an off-flavor described with 'burnt' and 'spicy' was detected. Gas chromatography-olfactometry (GCO) and gas chromatography-mass spectrometry (HRGC-MS) revealed 3-hydroxy-4,5-dimethyl-2[5H]-furanone (sotolon) as the off-flavor formed during storage. Among the ingredients of the soft drinks, ethanol and ascorbic acid were found to be the essential precursors of sotolon. Two formation pathways were postulated on the basis of studies using ^{2}H- (D) and ^{13}C-labeled ethanol and ascorbic acid. Sotolon is either formed from two molecules of ethanol and carbons 2 and 3 of ascorbic acid (pathway 1) or it is generated from one molecule of ethanol and carbons 3, 4, 5 and 6 of ascorbic acid (pathway 2).

3-Hydroxy-4,5-dimethyl-2[5H]-furanone (sotolon) contributes significantly to the characteristic impression of several foods, e.g., stewed beef (*1*), roasted coffee (*2*, *3*), bread crust (*4*), as well as flor sherry and botrytized wine (*5*, *6*). It is a potent flavor volatile with a very low odor threshold value of 0.02 ng/L air (*7*).

Sotolon has been described as a degradation product of glutamic acid and threonine (Fig. 1) (*8*). Later, the formation of sotolon in sake and wine has been reported (*9*). Enzymatically catalyzed as well as thermally induced oxidative deamination of 4-hydroxyisoleucine yielded also sotolon (*7*, *10*, *11*) and, recently it was detected after heating an aqueous solution containing hydroxyacetaldehyde and butan-2,3-dione at pH 5 (*12*).

The aim of our study was the identification of an off-flavor, described as 'burnt' and 'spicy' which was detected during storage of soft drinks. The beverages were composed of water and ethanolic citrus essences supplemented with ascorbic acid as antioxidant.

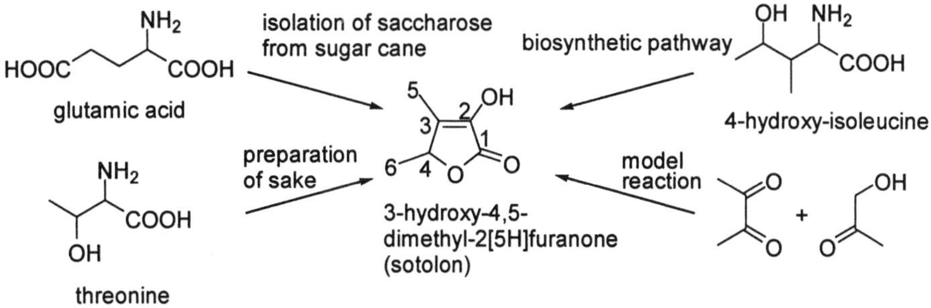

Figure 1. Formation pathways of 3-hydroxy-4,5-dimethyl-2[5H]furanone (sotolon).

Experimental

Preparation of model soft drinks. Model soft drinks were prepared from the individual essential oils (lemon, lime), citral or individual essences and with/without the addition of ascorbic acid according to a recipe provided by Doehler, Darmstadt. The model soft drinks were stored at 70 °C for two weeks and analyzed by HRGC-MS.

Model solutions. Solutions consisting of 250 µL of ethanol and 83 mg of ascorbic acid and 42 mL of water were stored for two weeks at 70 °C. The solutions were extracted and subsequently analyzed by HRGC-MS. Several modifications were applied: i) addition of EDTA (50 mg), ii) storage under nitrogen, iii) substitution of ascorbic acid by dehydroascorbic acid.

Extraction. Soft drinks (1000 mL), model soft drinks (250 mL), and model solutions (42 mL) were subjected to continuous liquid-liquid extraction using 250 mL of pentane-CH_2Cl_2 2:1 for 24 hrs. The organic extract was dried (Na_2SO_4), and concentrated (about 0.5 mL). The extract was analyzed by HRGC-MS analysis.

High-Resolution Gas Chromatography-Mass Spectrometry (HRGC-MS). HRGC-MS-analysis was performed with a Fisons GC 8000 coupled with a Fisons Instruments MD800 quadrupole-mass detector fitted with a split-injector (1:20) at 230 °C. A J &W DB-Wax fused silica capillary column (30 m x 0.25 mm i.d.; d_f = 0.25 µm) which was programmed from 50 °C for 3 min, then to 220 °C (for 10 min) at 4 °C/min was used with 2 mL/min of helium gas. Significant MS operating parameters: ionization voltage, 70 eV (electron impact ionization); ion source temperature, 220 °C; interface temperature, 250 °C; scan range 40-250 u; scan duration, 0.69 s. Constituents were identified by comparison of their mass spectra and retention indices with those of authentic reference compounds.

High-Resolution Gas Chromatography-Olfactometry Analysis (GCO). The

HRGC-olfactometry analysis was performed with a Dani 6500 HRGC-system equipped with a special nose adapter as detector. The chromatographically separated compounds were sensorically characterised by the operator using the nose adapter and the retention times were noticed. The chromatographic condition were essentially the same as for HRGC-MS analysis.

Results and Discussion

During storage of citrus soft drinks supplemented with vitamins an off-flavor described as 'burnt' and 'spicy' was observed. In extracts obtained by liquid-liquid extraction of the stored citrus soft drinks a trace constituent of the extract showed the same odor expression as the off-flavor exhibited by the soft drink. This compound was identified as 3-hydroxy-4,5-dimethyl-2[5H]-furanone (sotolon) (Fig. 2) by HRGC-MS analysis and by comparison with the authentic reference compound.

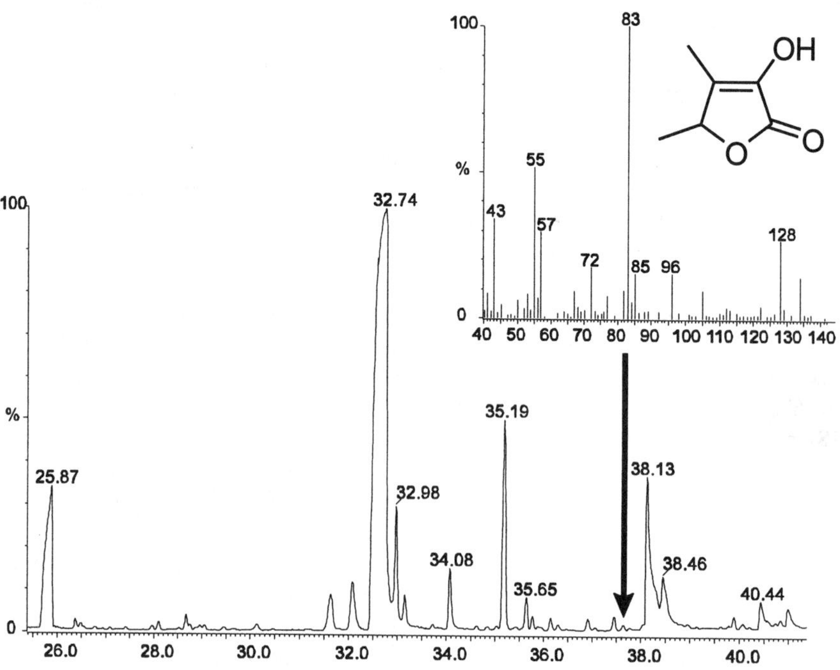

Figure 2. HRGC-MS analysis of an extract obtained by liquid-liquid extraction from a citrus soft drink exhibiting an off-flavor.

To gain insight into the formation of sotolon we prepared several model soft drinks composed of the individual essential oils or individual essences (ethanolic extracts of essential oils) and with/without the addition of ascorbic acid. Model soft drinks were stored at 70 °C for two weeks and they were analyzed by HRGC-MS. The typical off-flavor sotolon was only detected in model soft drinks prepared with essences and supplemented with ascorbic acid. Model soft drinks composed of the individual essential oils and ascorbic acid were devoid of the off-flavor.

As the essences are ethanolic extracts of essential oils, the major difference is the occurrence of ethanol in the essences. Therefore, we assumed that ethanol and ascorbic acid are important constituents of the soft drinks responsible for the formation of sotolon. The furanone was not formed in model soft drinks prepared with methanolic or propanolic extracts of essential oils.

The final evidence was given by a model solution consisting of only ethanol and ascorbic acid. After storage at 70 °C for two weeks this solution showed the typical sotolon off-flavor and sotolon was detected by HRGC-MS. Storage of the same solution under nitrogen or addition of EDTA inhibited the formation of the furanone, indicating the importance of oxygen and metal ions such as iron, respectively for the production of the malodor. Sotolon was also produced from dehydroascorbic acid and ethanol. Gas chromatographic resolution of the enantiomers of sotolon on a chiral phase showed the formation of racemic sotolon during the storage of the model solutions. Quantitative analysis revealed that the amount of sotolon gradually increased during a storage period of 3 weeks.

The formation of sotolon from ethanol and ascorbic acid was further investigated in model reactions with isotopically labeled precursors. 1,1-D_2-ethanol, 2,2,2-D_3-ethanol, and 1,1,2,2,2-D_5-ethanol were added to aqueous solutions of unlabeled ascorbic acid and the mixtures were stored at 70 °C for two weeks. HRGC-MS analysis revealed that sotolon formed from 1,1-D_2-ethanol carried one deuterium atom. The deuterium atom might be attached to carbon 4 of the sotolon molecule as this is the only carbon carrying one proton (Fig. 3).

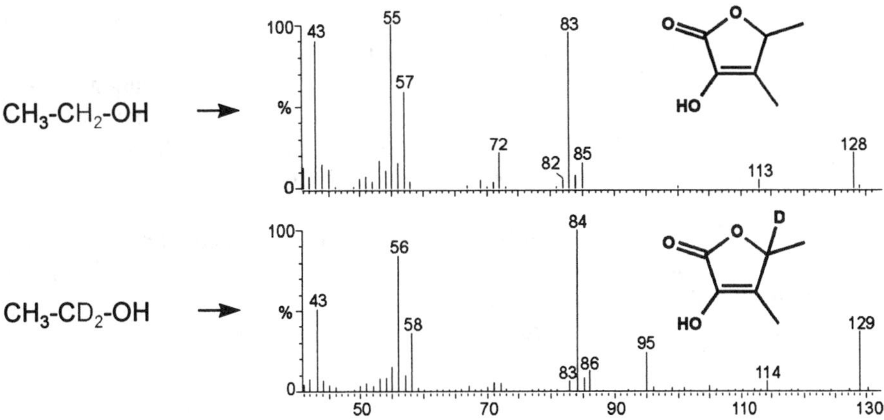

Figure 3. Mass spectrum of sotolon formed from ascorbic acid and ethanol (upper trace) and sotolon formed from ascorbic acid and 1,1-D_2-ethanol (lower trace).

Two isotopomers of sotolon were detected by HRGC-MS analysis in the model reaction containing 2,2,2-D_3-ethanol (Fig. 4). The mass spectra showed that the first isotopomer contained six and the second carried three deuterium atoms. The deuterium atoms in the first isotopomer are located in both methyl groups while only one methyl group contained the deuterium atoms of the second isotopomer. Therefore, two formation pathways for the furanone might exist.

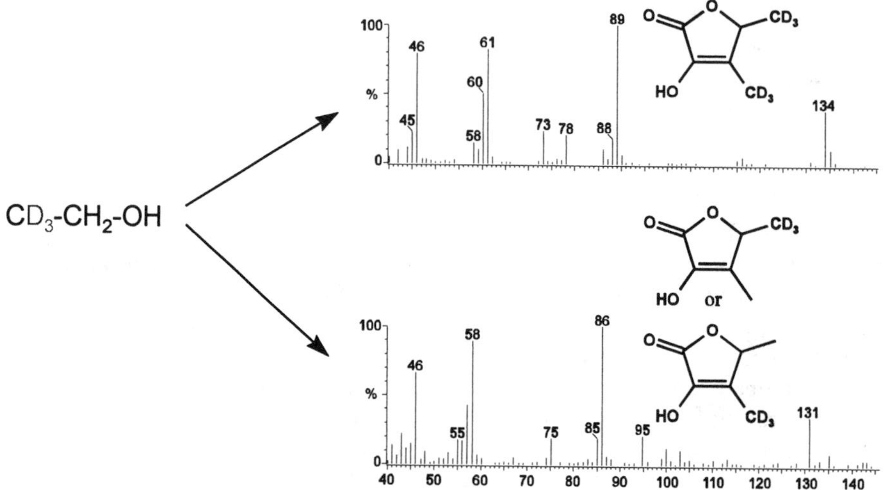

Figure 4. Mass spectra of sotolon isotopomer 1 (upper trace) and isotopomer 2 (lower trace) formed from 2,2,2-D_3-ethanol and ascorbic acid.

In order to localize the deuterium label in one of the two methyl groups of the second isotopomer we analyzed solutions containing unlabeled ascorbic acid and 1,1,2,2,2-D_5-ethanol. HRGC-MS separation showed again the formation of two isotopomers of sotolon one carrying seven and the other carrying four deuterium atoms. In the case of the first isotopomer all protons are replaced by deuterium atoms. The four deuterium atoms of the second isotopomer are located in the methyl group 6 attached to carbon 4 and at carbon 4 (Fig. 1).

Further evidence was provided by HRGC-MS analysis of sotolon produced from $^{13}C_2$-ethanol and unlabeled ascorbic acid (Fig. 5). Two isotopomers were generated, but resolution of the ^{13}C-labled compounds by HRGC was not achieved. Thus, the mass spectrum shown in Fig 5 represents the sum of two spectra.

$^{13}CH_3$-$^{13}CH_2$-OH

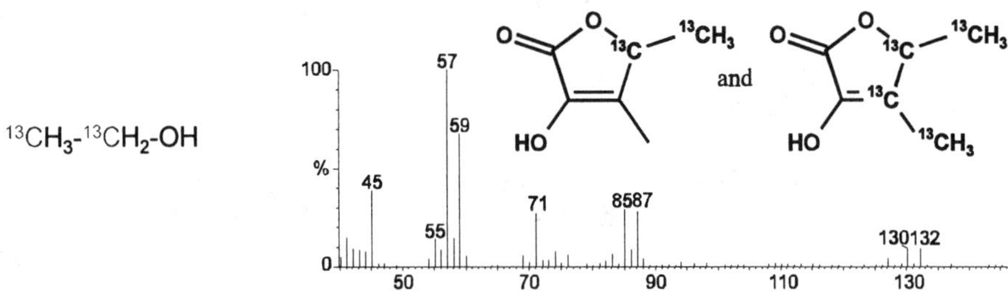

Figure 5. Sotolon formed from $^{13}C_2$-ethanol and unlabeled ascorbic acid.

Two formation pathways for sotolon exist. In pathway 1 two molecules of ethanol are incorporated into the sotolon molecule whereas in pathway 2 only one molecule of ethanol is transformed to sotolon. Consequently, two respectively four carbons of the formed sotolon molecule originate from ascorbic acid.

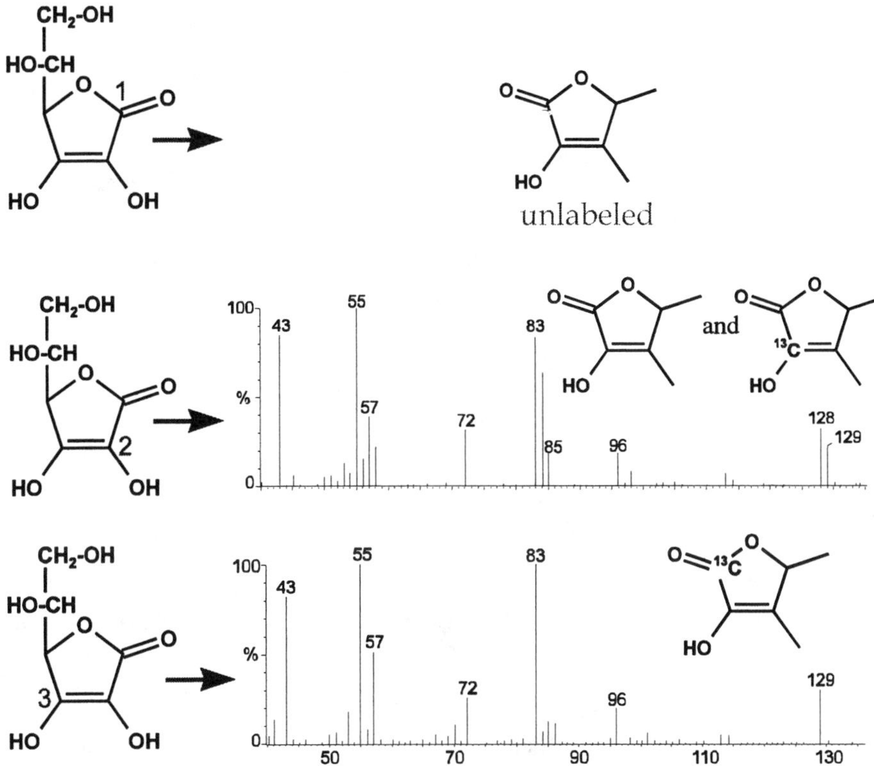

Figure 6. Sotolon formed from 1-^{13}C-, 2-^{13}C-, and 3-^{13}C-ascorbic acid and unlabeled ethanol.

Hence, we performed model reactions using solutions of 1-^{13}C-, 2-^{13}C-, and 3-^{13}C-ascorbic acid as well as unlabeled ethanol in order to identify the carbons originating from ascorbic acid (Fig. 6). Unlabeled sotolon was obtained from 1-^{13}C-ascorbic acid. Therefore, carbon 1 of the acid is not involved in the formation of the furanone. It is probably lost as CO_2 by decarboxylation as recently described (*13*). Unlabeled and singly labeled sotolon was produced by 2-^{13}C-ascorbic acid and only singly labeled sotolon was formed by 3-^{13}C-ascorbic acid (Fig. 6).

This results can be interpreted with the conclusion drawn from the studies with labeled ethanols. Carbon 4 and its attached methyl group is always generated by ethanol. Carbon 3 and its attached methyl group originates either from ethanol or ascorbic acid. The remaining carbons of the sotolon molecule must be provided by ascorbic acid since there is no further carbon source available. However, carbon 1 of the ascorbic acid molecule is not involved. Summarizing the results, two formation pathways can be postulated (Fig. 7). Sotolon is either formed from two molecules of ethanol and carbons 2 and 3 of ascorbic acid (pathway 1) or it is generated from one molecule of ethanol and carbons 3, 4, 5, and 6 of ascorbic acid (pathway 2).

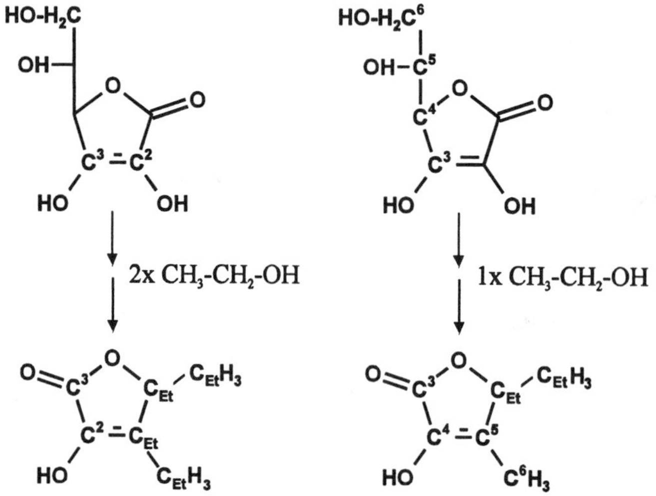

HO-H$_2$C | OH | C^3=C^2 | HO | OH | O | =O

HO-H$_2$C^6 | OH-C^5 | C^4 | C^3= | HO | OH | O | =O

2x CH$_3$-CH$_2$-OH

1x CH$_3$-CH$_2$-OH

O=C^3 | O | C$_{Et}$ | C$_{Et}$H$_3$ | C^2=C$_{Et}$ | HO | C$_{Et}$H$_3$

O=C^3 | O | C$_{Et}$ | C$_{Et}$H$_3$ | C^4=C^5 | HO | C^{6}H$_3$

Figure 7. Proposed formation pathways of sotolon from ascorbic acid and ethanol.

By comparison of the mass spectra given in Fig. 6 it is even possible to locate the fragments of the ascorbic acid in the formed sotolon molecule. In the mass spectrum of the furanone formed from 3-^{13}C-ascorbic acid only the molecular ion $[M]^+$ contained a ^{13}C isotope (*m/z* 129) (Fig. 6). The rest of the mass spectrum is similar to that of unlabeled sotolon. The fragment ion *m/z* 83 resulting by the loss of -CHO_2 (typical for lactones) carries no label. This implies that the ^{13}C-label is located in carbon 1 of the sotolon molecule. The mass spectrum of sotolon formed by 2-^{13}C-ascorbic acid also showed a labeled molecule ion $[M]^+$ (*m/z* 129). But after loss of

$-CHO_2$ the resulting fragment still carries ^{13}C (*m/z* 84). (Fig. 6). Following pathway 1 in Fig. 7 this means that carbon 2 of the ascorbic acid molecule forms carbon 2 of the sotolon molecule.

The theory was tested with model solutions containing 2-^{13}C-, respectively 3-^{13}C-ascorbic acid and 1,1,2,2,2-D_5-ethanol. Two isotopomers of sotolon (*m/z* 132 and *m/z* 136) were produced from 2-^{13}C-ascorbic acid and D_5-ethanol. The furanone with *m/z* 132 is composed of one molecule D_5-ethanol and the carbons 3, 4, 5, and 6 of the 2-^{13}C-ascorbic acid molecule. The other isotopomer is formed from two molecules D_5-ethanol and carbons 2 and 3 of the 2-^{13}C-ascorbic acid molecule. Two isotopomers of sotolon (*m/z* 133 and *m/z* 136) were formed from 3-^{13}C-ascorbic acid and D_5-ethanol. Furanone (*m/z* 136) contained two molecules of D_5-ethanol and the carbons 2 and 3 of 3-^{13}C-ascorbic acid. Sotolon (*m/z* 133) was composed of one D_5-ethanol molecule and the carbon 3, 4, 5, and 6 of 3-^{13}C-ascorbic acid.

The results demonstrate the formation of sotolon by ascorbic acid and ethanol and provide evidence for two different pathways leading to the off-flavor (Fig. 7). Although various beverages contain ethanol and ascorbic acid the off-flavor caused by sotolon has been described so far only in citrus soft drinks. It is possible that natural ingredients in beverages other than citrus soft drinks bind reactive intermediates of the sotolon pathways and therefore inhibit the formation of the off-flavor.

References

1. Guichard, E.; Etiévant, P.; Henry, R.; Mosandl, A. *Z. Lebensm. Unters. Forsch.* **1992**, *195*, 540-544.
2. Blank, I.; Sen, A.; Grosch, W. *Z. Lebensm. Unters. Forsch.* **1992**, *195*, 239-245.
3. Semmelroch, P.; Laskawy, G.; Blank, I.; Grosch, W. *Flav. Frag. J.* **1995**, *10*, 1-7.
4. Schieberle, P.; Grosch, W. *Z. Lebensm. Unters. Forsch.* **1994**, *198*, 292-296.
5. Guichard, E.; Etiévant, P.; Henry, R.; Mosandl, A. *Z. Lebensm. Unters. Forsch.* **1992**, *195*, 540-544.
6. Martin, B.; Etiévant, P. X.; Le Quéré, J. L.; Schlich, P. *J. Agric. Food Chem.* **1992**, *40*, 475-478.
7. Blank, I.; Lin, J.; Fumeaux, R.; Welti, D. H.; Laurent, B. *J. Agric. Food Chem.* **1996**, *44*, 1851-1856.
8. Sulser, H.; De Pizzol, J.; Büchi, W. *J. Food Sci.* **1967**, *32*, 611-615.
9. Kobayashi, A. In: *Flavor Chemistry Trends and Developments*; Teranishi, R.; Buttery, R. G.; Shahidi, F., Eds.; American Chemical Society: Washington, DC, 1989, p 49-59.
10. Sauvaire, Y.; Givardon, P.; Baccou, J. C.; Risterucii, A. M. *Phytochemistry* **1984**, *23*, 479-486.
11. Blank, I.; Schieberle, P.; Grosch, W. In: *Progress in Flavor Precursor Studies*; Schreier, P.; Winterhalter, P., Eds.; Allured: Carol Stream, 1992, p 103-109.
12. Hofmann, T.; Schieberle, P. In: *Flavor Science - Recent Developments*; The Royal Society of Chemistry: London, 1996, p 182-187.
13. Shin, D. B.; Feather, M. S. *J. Carbohydr. Chem.* **1990**, *9*, 461-469.

Additional Properties of Flavor Components and Flavors

Chapter 19

Antimicrobial Activities of Isothiocyanates

Hideki Masuda[1], Yasuhiro Harada[1], Noriaki Kishimoto[2], and Tatsuo Tano[2]

[1]Material R&D Laboratories, Ogawa and Company, Ltd., 1–2 Taiheidai, Shoo-cho, Katsuta-gun, Okayama 709–4321, Japan
[2]Faculty of Food Culture, Kurashiki Sakuyo University 3515, Nagao, Tamashima, Kurashiki-city, Okayama 710–0251, Japan

Wasabi (Japanese horseradish), horseradish, and mustard are known to have remarkable antimicrobial activity. The study of the antimicrobial activity is important because of its application for food preservation. The antimicrobial activity of allyl isothiocyanate, the main volatile component of wasabi, horseradish, and mustard, has been reported in detail. In this study, the effect of a variety of volatile isothiocyanates against eleven microorganisms was evaluated. In addition, in order to evaluate the antimicrobial activities of the isothiocyanates, those of catechin, its analogues, and food preservatives have been studied.

The development of food preservation is important for the supply of safe and high-quality food. At present, for example, heat sterilization, the control of water activity and pH value, cold storage, skin packaging, and food preservatives are used as methods of food preservation. Many kinds of spices and herbs that play an important role in flavoring are also used for food preservation (*1-4*). Taking into account the flavor deterioration by sterilization, the antimicrobial effect of spices is considered to be very important.

It is well-known that extracts of wasabi, horseradish, and mustard have remarkable antimicrobial activity (*5*, *6*). In particular, the marked antimicrobial activity of allyl isothiocyanate, the main volatile component of wasabi, horseradish, and mustard, has been studied in detail (*7-11*). However, the effects of other volatile isothiocyanates on microorganisms have so far been little reported (*12-16*).

This study focuses on the antimicrobial activities of twenty kinds of isothiocyanates against five species of bacteria and two species of fungi. In addition, the antimicrobial properties of streptomycin sulfate (an antibiotic), (+)-catechin, (-)-epicatechin, (-)-epigallocatechin, (-)-epicatechin gallate, and (-)-epigallocatechin gallate, i.e., nonvolatile components in green tea extract, and seven food preservatives have been studied in order to compare with those of the isothiocyanates. The effect of the isothiocyanates against four species of lactic acid bacteria which are used for the processing of many dairy products has also been investigated.

Experimental

Four ω-alkenyl isothiocyanates (except for allyl isothiocyanate) were prepared by the isomerization of the corresponding ω-alkenyl thiocyanates (*17*). Five ω-alkenyl isothiocyanates were converted to the corresponding ω-methylthioalkyl isothiocyanates (*18*). Five ω-methylthioalkyl isothiocyanates were oxidized to the corresponding ω-methylsulfinylalkyl isothiocyanates using m-chloroperbenzoic acid. The other isothiocyanates, (+)-catechin, (-)-epicatechin, (-)-epigallocatechin, (-)-epicatechin gallate, (-)-epigallocatechin gallate, ethyl p-hydroxybenzoate, n-propyl p-hydroxybenzoate, n-butyl p-hydroxybenzoate, propionic acid, potassium sorbate, sodium benzoate, and sodium dehydroacetate were purchased from commercial sources.

Bacillus subtilis IFO 3134,, methicillin-sensitive *Staphylococcus aureus* IFO 12732, methicillin-resistant *Staphylococcus aureus* IFO 12732, *Escherichia coli* IFO 3301, *Pseudomonas aeruginosa* IFO 3080, *Lactobacillus casei* subsp. *casei* JCM 1134, *Lactobacillus helveticus* JCM 1120, *Lactobacillus lactis* subsp. *cremoris* IAM 1150, and *Streptococcus salvarius* subsp. *thermophilus* IAM 10064 were obtained from the Institute for Fermentation, Osaka (Osaka). The PYMG culture medium was made from peptone (5 g), yeast extract (1.5 g), meat extract (1.5 g), glucose (1 g), sodium chloride (3.5 g), disodium hydrogen phosphate (3 g), potassium dihydrogen phosphate (1.32 g), and distilled water (1000 mL). The pH was adjusted to 7.0-7.1. The GWYP culture medium was made from peptone (5 g), yeast extract (1.5 g), glucose (1 g), magnesium sulfate hydrate (0.2 g), and whey (1000 mL). The pH was adjusted to 6.5.

Candida albicans IFO 1385 and *Aspergillus niger* IFO 4414 were obtained from the Institute for Fermentation, Osaka (Osaka). The culture medium was made from peptone (10 g), glucose (40 g), and distilled water (1000 mL). The pH was adjusted to 5.6±1.

Each isothiocyanate, streptomycin sulfate, each catechin and its analogue, or each food preservative in 80 % aqueous methanol solution was added to the test tube containing the culture medium. The initial concentrations of sample were varied between 1 and 1000 ppm by the continuous two fold dilution method. No addition of sample was regarded as the control. The test tube was incubated at 35°C for 6 h for *B. subtilis* IFO 3134, methicillin-sensitive *S. aureus* IFO 12732, methicillin-resistant *S.*

aureus IFO 12732, *E. coli* IFO 3301, *P. aeruginosa* IFO 3080, *L. casei* subsp. *casei* JCM 1134, *L. helveticus* JCM 1120, *L. lactis* subsp. *cremoris* IAM 1150, and *S. salvarius* subsp. *thermophilus* IAM 10064, at 25°C for 22 h for *C. albicans* IFO 1385, and at 25°C for 15 h for *A. niger* IFO 4414. The inhibitory concentration was confirmed by measuring the transmittance at 630 nm using a Bausch & Lomb Spectronic 20 spectrophotometer. All tests were run in triplicate and averaged.

Results and Discussion

The antimicrobial activities of the alkyl isothiocyanates (isopropyl- (**1**), *sec*-butyl- (**2**), and isobutyl- (**3**)), the ω-alkenyl isothiocyanates (allyl- (**4**), 3-butenyl- (**5**), 4-pentenyl- (**6**), 5-hexenyl- (**7**), and 6-heptenyl- (**8**)), the aryl isothiocyanates (benzyl- (**9**) and 2-phenethyl- (**10**)), the ω-methylthioalkyl isothiocyanates (3-methylthiopropyl- (**11**), 4-methylthiobutyl- (**12**), 5-methylthiopentyl- (**13**), 6-methylthiohexyl- (**14**), and 7-methylthioheptyl- (**15**)), and the ω-methylsulfinylalkyl isothiocyanates (3-methylsulfinylpropyl- (**16**), 4-methylsulfinylbutyl- (**17**), 5-methylsulfinylpentyl- (**18**), 6-methylsulfinylhexyl- (**19**), and 7-methylsulfinylheptyl- (**20**)) in PYMG culture medium are shown in Figure 1. The antimicrobial activities of the alkyl isothiocyanates **1**, **2**, and **3** were found to be low against the gram-positive bacteria, i.e., *B. subtilis*, *MSSA* (methicillin-sensitive *S. aureus*), and *MRSA* (methicillin-resistant *S. aureus*), the gram-negative bacteria, i.e., *E. coli* and *P. aeruginosa*, and the fungi, i.e., *C. albicans* and *A. niger*, compared with those of the other isothiocyanates. The ω-alkenyl isothiocyanates **4-8**, in general, showed high antimicrobial activities. The antimicrobial activities against *C. albicans* and *A. niger* were extremely high.

The aryl isothiocyanates **9** and **10**, in general, showed high antimicrobial activities against all the microorganisms except for *P. aeruginosa*, compared with the other isothiocyanates. Taking into account that the aryl isothiocyanates **9** and **10** are the characteristic volatile components in horseradish, wide application of horseradish to food preservation will be expected (*19*). In addition, it is interesting that the low MIC (minimum inhibitory concentration) values of benzyl isothiocyanate (**9**) against the bacteria were similar to those of streptomycin sulfate (The MIC values of streptomycin sulfate against *B. subtilis*, *MSSA*, *MRSA*, *E. coli*, and *P. aeruginosa* were 1, 2, 2, 4, and 4 ppm, respectively. The MIC values of streptomycin sulfate against *C. albicans* and *A. niger* were both more than 1000 ppm). Due to its pronounced high activity against *MRSA*, which is an important source of internal infection in hospitals, **9** is expected to be used widely in unsanitary places. The antimicrobial activities of the ω-methylthioalkyl isothiocyanates **11-15** against the gram-positive bacteria, i.e., *B. subtilis*, *MSSA*, *MRSA*, and fungi, i.e., *C. albicans* and *A. niger*, were higher than those against the gram-negative bacteria, i.e., *E. coli* and *P. aeruginosa*. The ω-methylsufinylalkyl isothiocyanates **16-20** had high antimicrobial activities against gram-positive bacteria and *A. niger*. However, in general, the antimicrobial activities against the gram-negative bacteria and *C. albicans* were found to be lower.

a

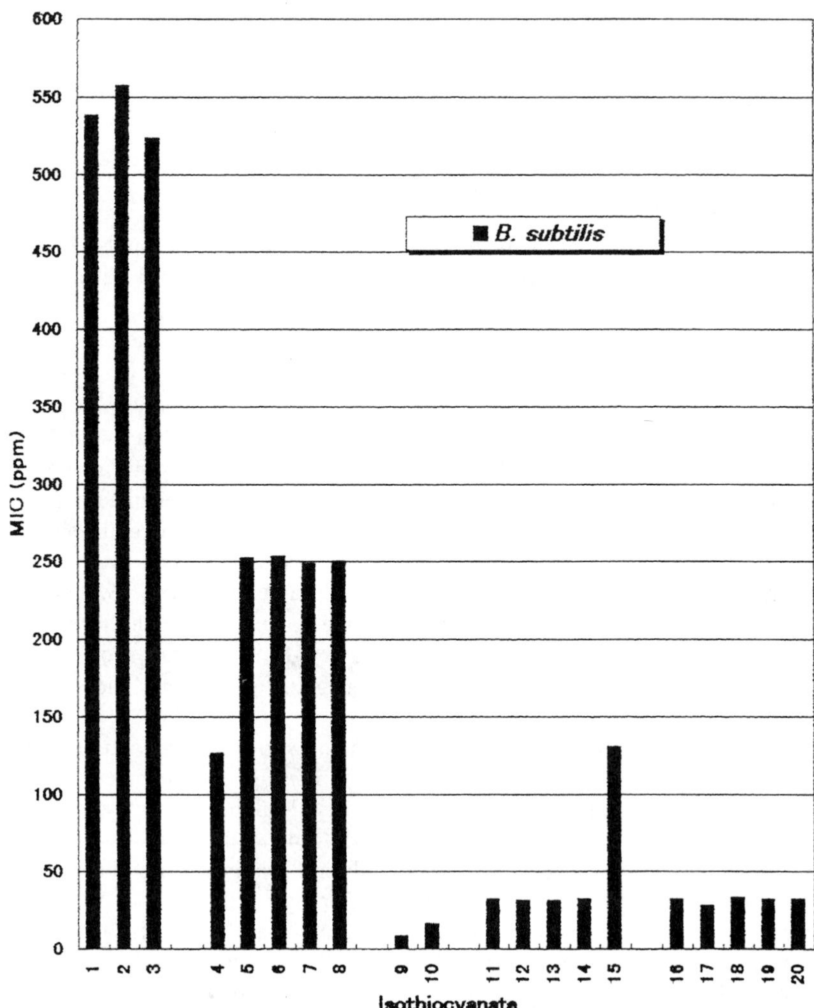

Figure 1. The MIC (minimum inhibitory concentration) values of the isothiocyanates against B. subtilis *(a), MSSA (methicillin-sensitive* S. Aureus *and MRSA (methicillin-resistant* S. aureus *(b),*E. Coli *(c), and* P. aeruginosa *(d) in PYMG culture medium. The MIC values of the isothiocyanates against* C. albicans *(e) and* A. niger *(f).*

b

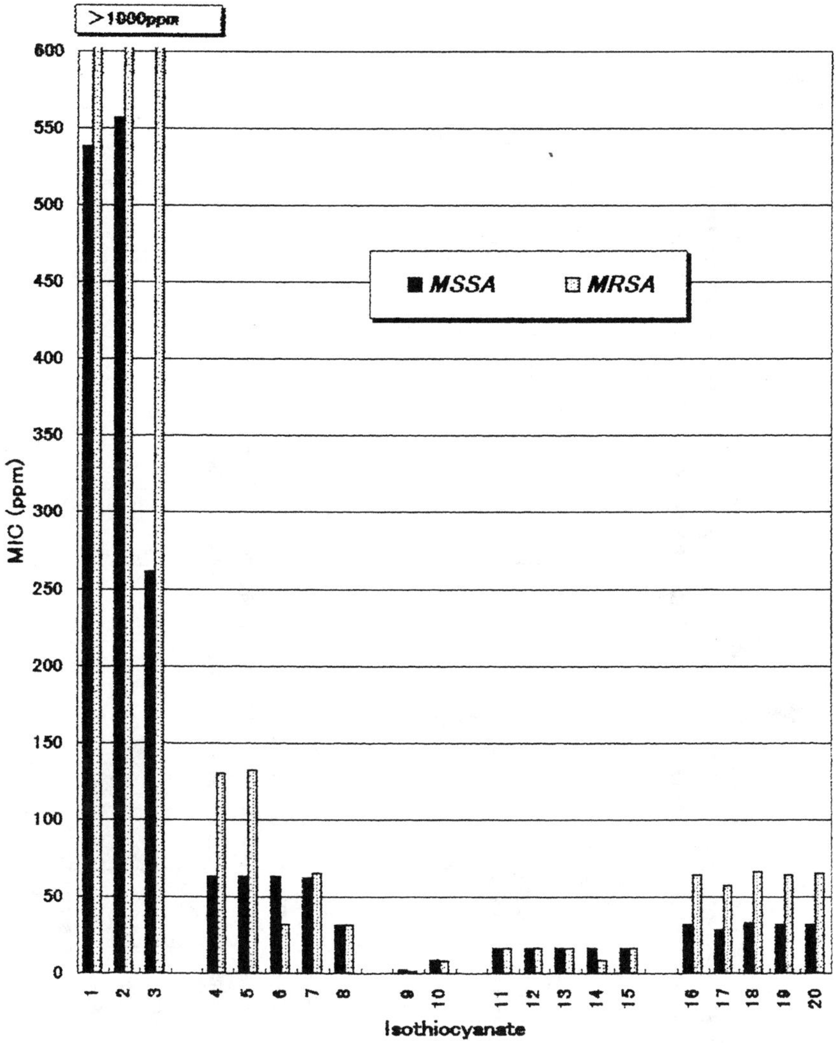

Figure 1. *Continued.*

Continued on next page.

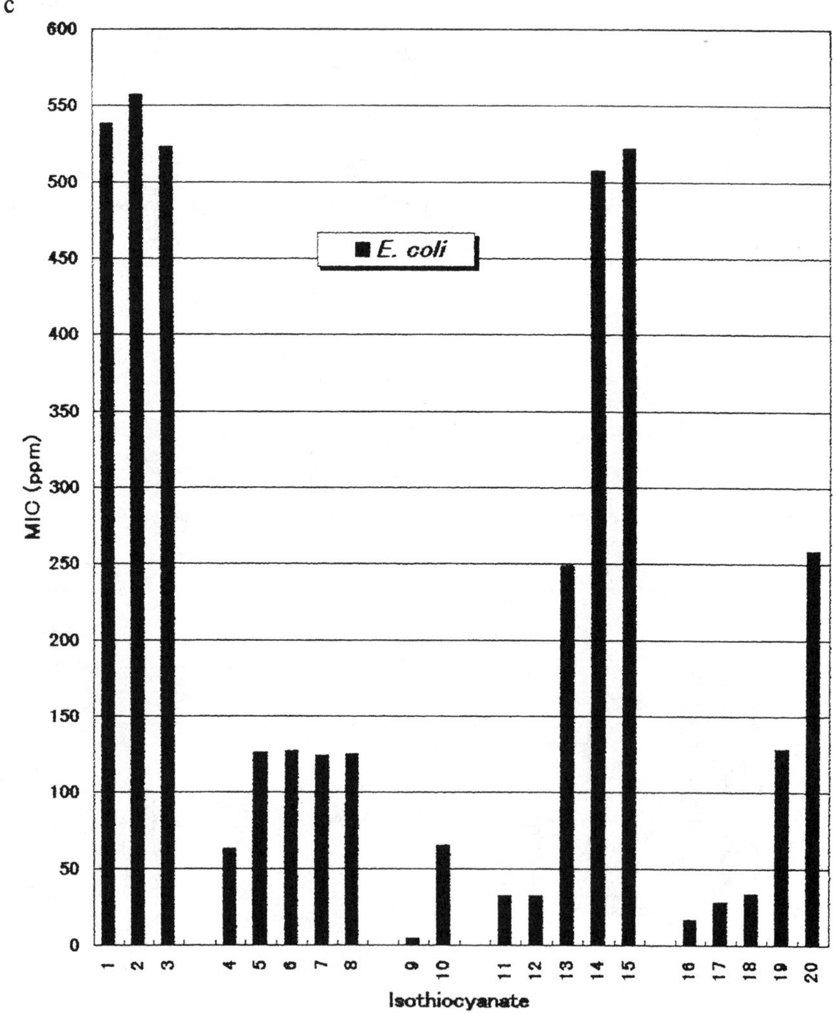

Figure 1. *Continued.*

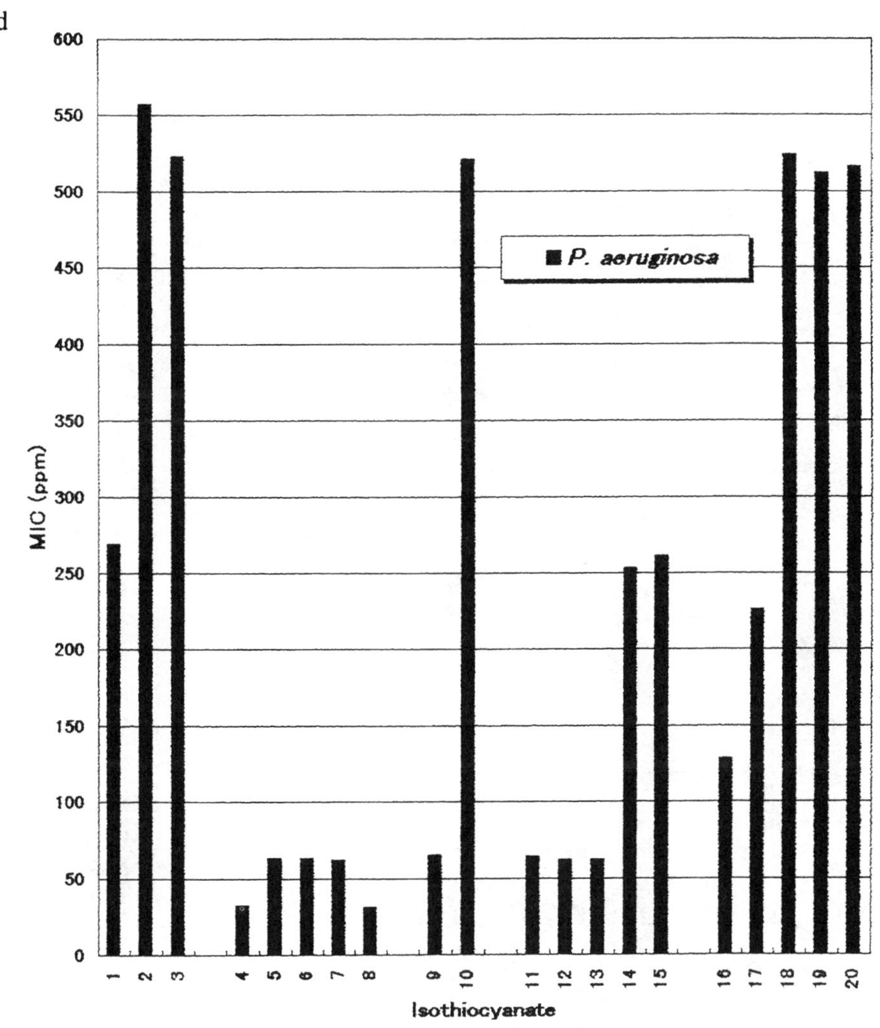

Figure 1. *Continued.*

Continued on next page.

e

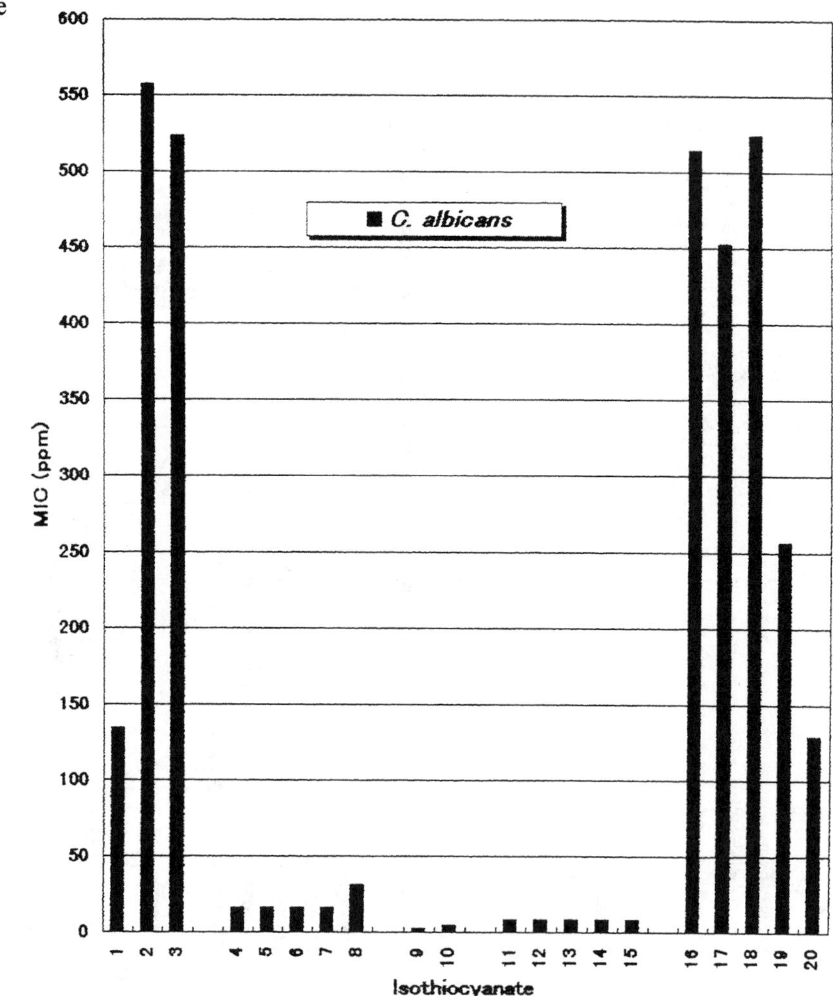

Figure 1. *Continued.*

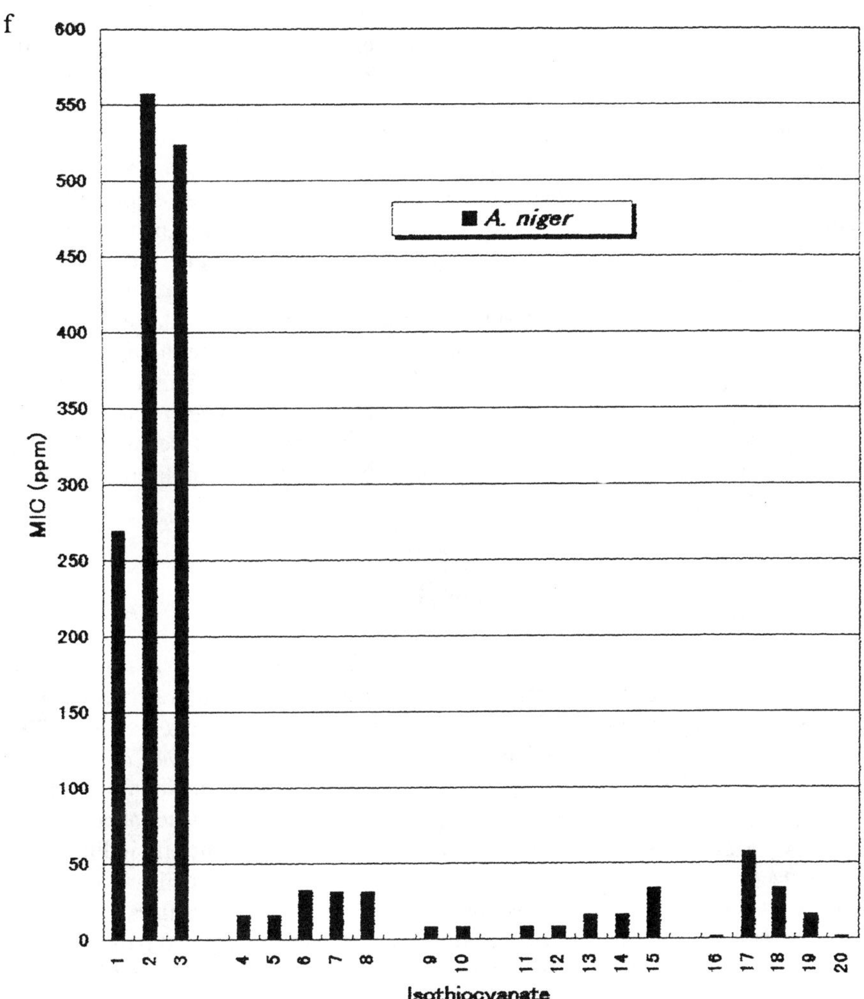

Figure 1. *Continued.*

preservation of food are shown in Figure 3. In general, the antimicrobial activities of the synthetic preservatives were lower than those of the isothiocyanates. Therefore, it is suggested that the isothiocyanates have significant usefulness compared with the above-mentioned preservatives. However, taking into account the stability and odor characteristics of the isothiocyanates in aqueous solution, it is important to decide the purpose of preservation, the preservation term, the form, the pH value, and the water activity of food (*24*).

The MIC values of the isothiocyanates against lactic acid bacteria, i.e., *L. casei*, *L. helveticus*, *L. lactis*, and *S. thermophilus*, in GWYP culture medium are shown in Figure 4. The GWYP culture medium was used because it is more appropriate for cultivation of lactic acid bacteria. The MIC values of the isothiocyanates against *B. subtilis*, *S. aureus*, *E. coli*, and *P. aeruginosa* measured in the same GWYP culture medium are shown in Figure 5. In general, the antimicrobial activities of the isothiocyanates against four lactic acid bacteria were lower than those against *B. subtilis*, *S. aureus*, *E. coli*, and *P. aeruginosa*. The lactic acid bacteria are known for use in producing a variety of fermented foods (*25*). With the remarkably different antimicrobial activity between the lactic acid bacteria and the other bacteria, i.e., *B. subtilis*, *S. aureus*, *E. coli*, and *P. aeruginosa*, in mind, there seemed to be a possibility of the isothiocyanates coexisting in order to inhibit the proliferation of any toxic microorganisms.

Conclusions

In general, ω-alkenyl-, aryl-, ω-methylthioalkyl-, and ω-methylsulfinylalkyl isothiocyanates showed high antimicrobial activities against gram-positive bacteria, i.e., *B. subtilis*, *MSSA* (methicillin-sensitive *S. aureus*), and *MRSA* (methicillin-resistant *S. aureus*), gram-negative bacteria, i.e., *E. coli* and *P. aeruginosa*, and fungi, i.e., *C. albicans* and *A. niger*, compared with catechin and its analogues, and the synthetic preservatives. It was particularly impressive that the antibacterial activity of benzyl isothiocyanate was the same as that of streptomycin sulfate. In addition, the

Taking into account the relationship between the structure and antimicrobial activity of the isothiocyanates, the action of a variety of isothiocyanates against the microorganisms is assumed to be highly selective. That is to say, the type of functional group, i.e., ω-alkenyl, aryl, ω-methylthioalkyl, and ω-methylsulfinylalkyl, and side-chain length of the isothiocyanates are supposed to provide different inhibition mechanisms against each microorganism (*12, 20-23*).

In analogy with wasabi, horseradish, and mustard extract, green tea extract is also called a shelf-life elongation material, which is used as a short-time preservative of unstable foods for a few hours or a few days, determined by Japanese food sanitation law. The antimicrobial activities of catechin and its analogues were found to be lower than those of the isothiocyanates except for the alkyl isothiocyanates as shown in Figure 2.

The antimicrobial activities of synthetic preservatives for the long-term

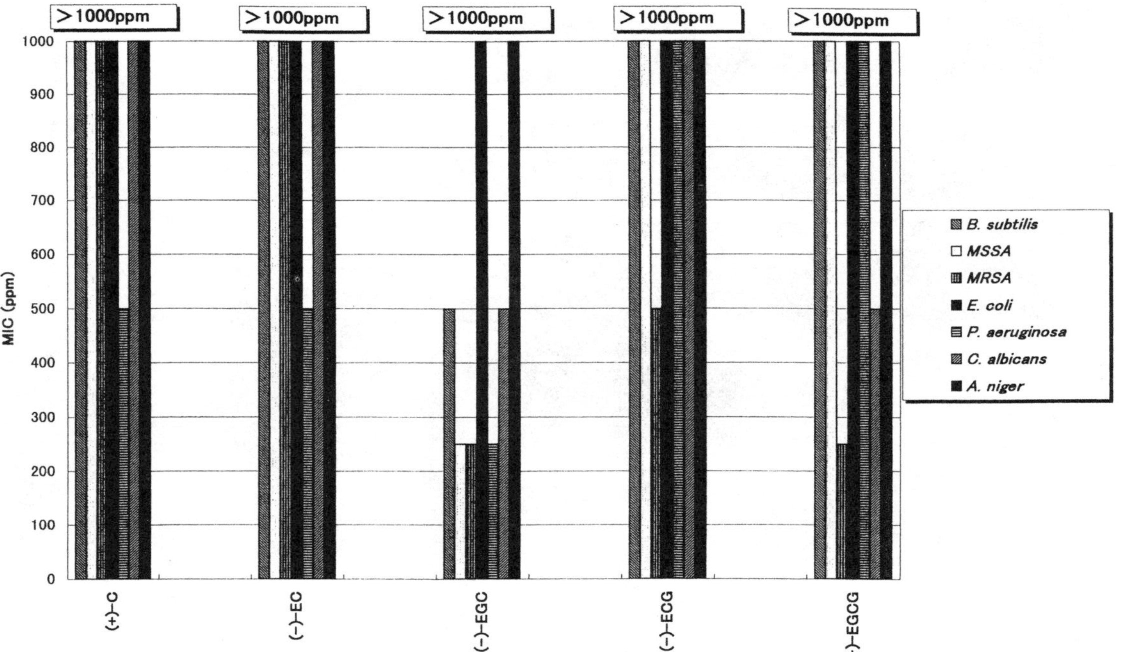

Figure 2. The MIC values of (+)-catechin ((+)-C), (–)-epicatechin ((–)-EC), (–)-epigallocatechin ((–)-EGC), epicatechin gallate ((–)-ECG), ((–)-epigallocatechin gallate ((–)-EGCG) against B. subtilis, *MSSA (methicillin-sensitive* S. aureus*), MRSA (methicillin-resistant* S. aureus*),* E. coli, *and* P. aeruginosa *in PYMG culture medium. The MIC values of (+)-C, (–)-EC, (–)-EGC, (–)-ECG, (–)-EGCG against* C. albicans *and* A. niger.

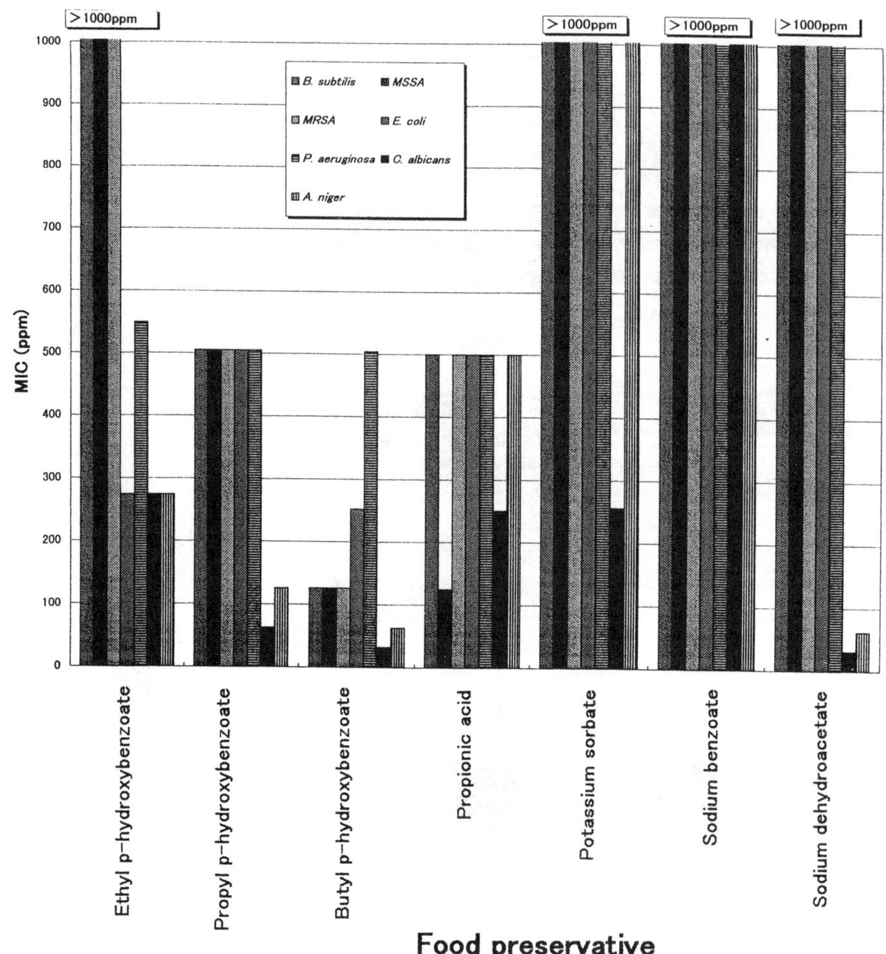

Figure 3. The MIC values of synthetic preservatives against B. subtilis, *MSSA, (methicillin-sensitive* S. aureus*), MRSA (methicillin-resistant* S. aureus*),* E. coli, *and* P. aeruginosa *in PYMG culture medium. The MIC values of the synthetic preservatives against* C. albicans *and* A. niger.

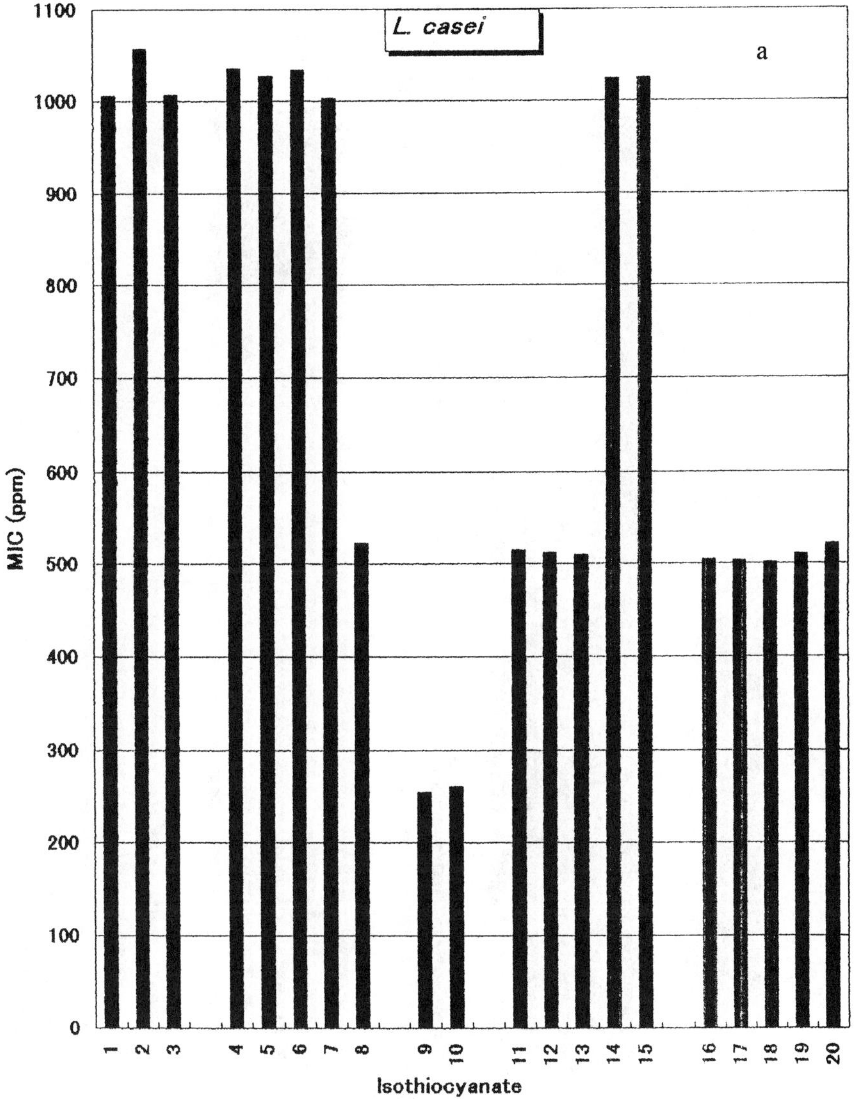

Figure 4. The MIC values of isothiocyanates against L. casei *(a),* L. helveticus *(b),* L. lactis *(c), and* S. thermophilus *(d) in GWYP culture medium.*

Continued on next page.

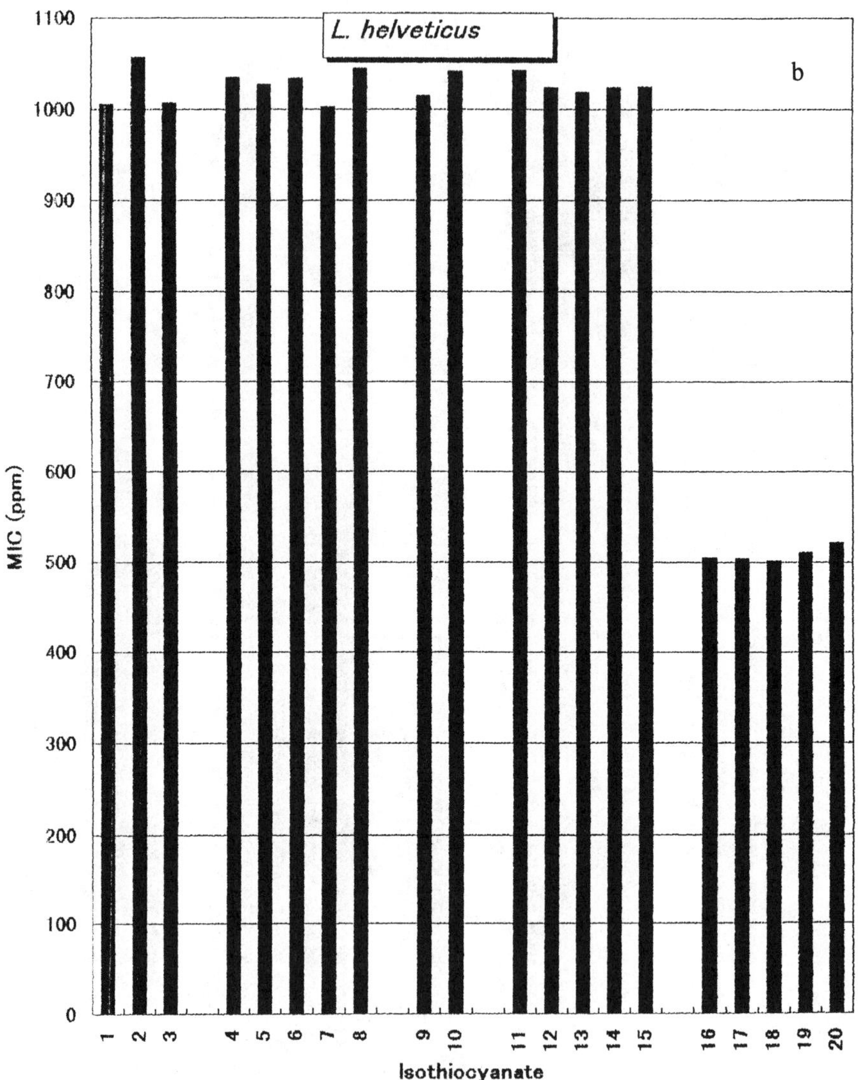

Figure 4. *Continued.*

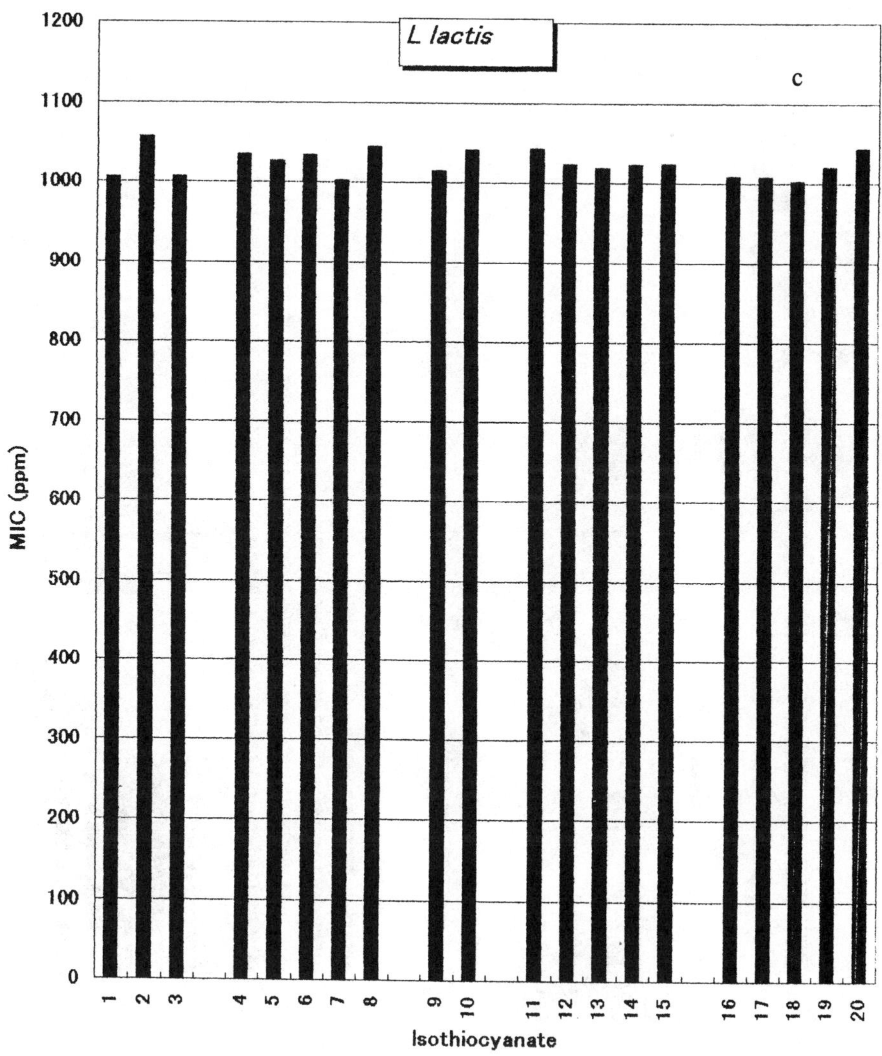

Figure 4. *Continued.*

Continued on next page.

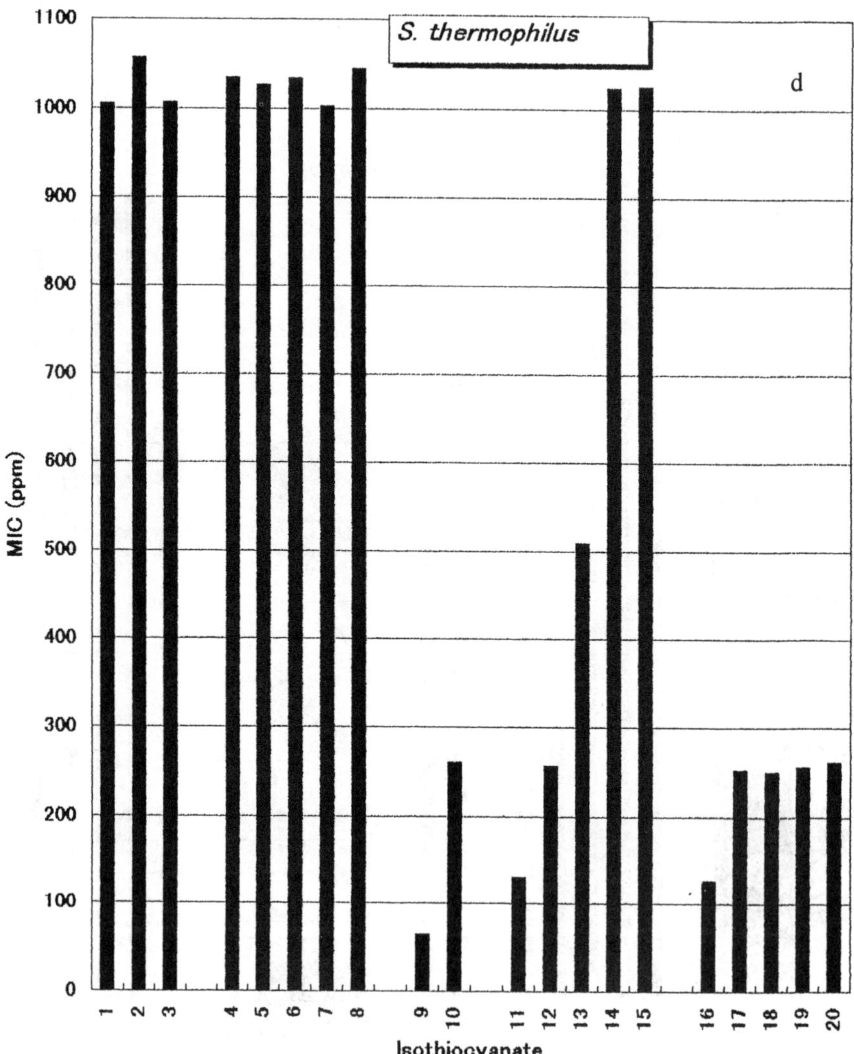

Figure 4. *Continued.*

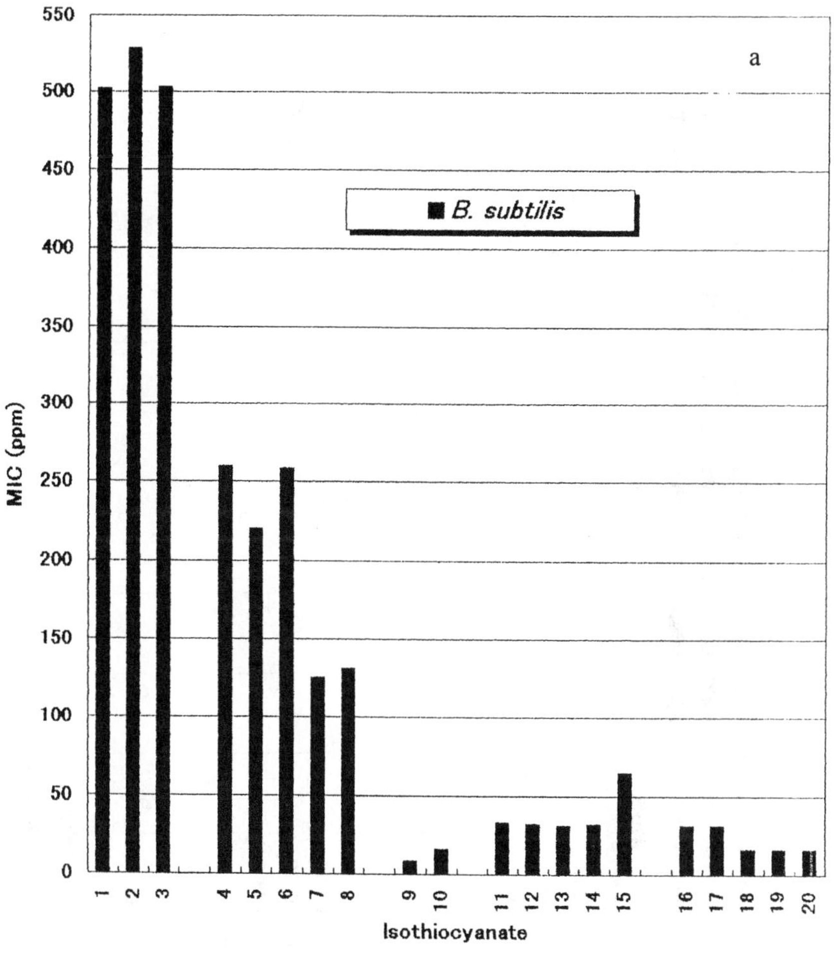

Figure 5. The MIC values of isothiocyanates against B. subtilis (a)*, MSSA, (methicillin-sensitive* S. aureus*) (b),* E. coli *(c), and* P. aeruginosa *(d) in GWYP culture medium.*

Continued on next page.

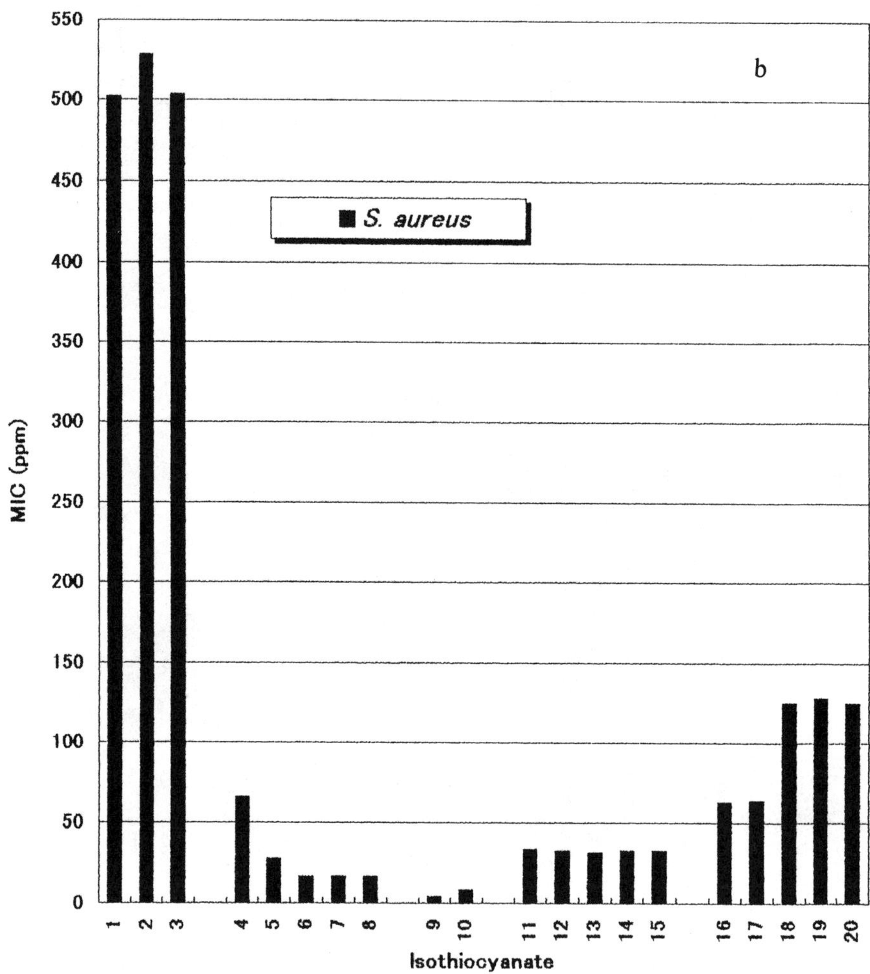

Figure 5. *Continued.*

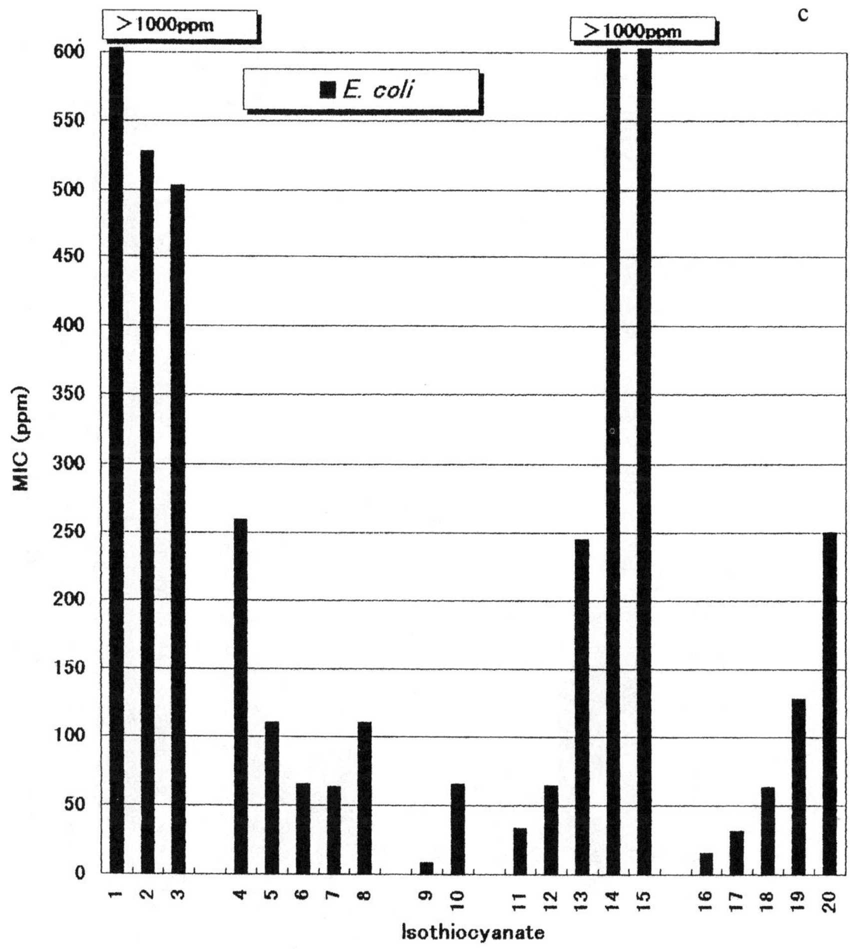

Figure 5. *Continued.*

Continued on next page.

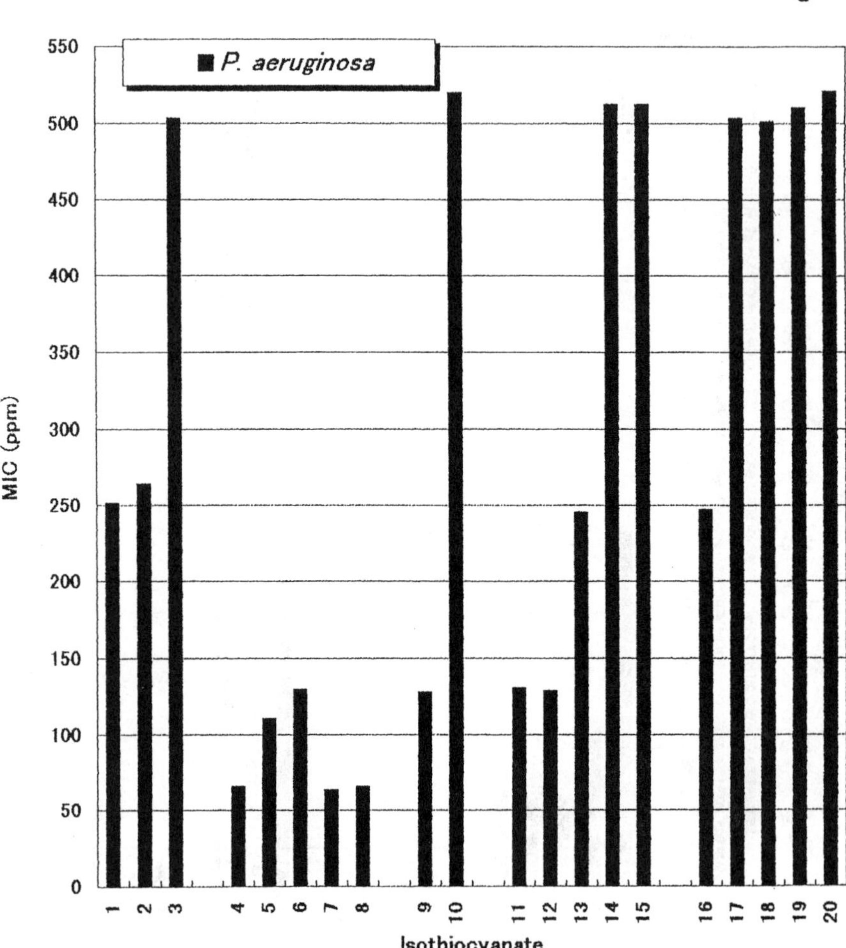

Figure 5. *Continued*

antimicrobial activities of the isothiocyanates against lactic acid bacteria were lower than those against the other bacteria, i.e., *B. subtilis*, *MSSA*, *E. coli*, and *P. aeruginosa*. Each isothiocyanate seemed to have a selective antimicrobial activity against a variety of microorganisms.

Literature Cited

1. Nakatani, N. In *Spices, Herbs, and Edible Fungi*; Geoge, C., Ed.; Developments in Food Science 34; Elsevier: Amsterdam, Netherlands, 1994; pp. 251-271.
2. Ueda, S.; Yamashita, H.; Nakajima, M.; Kuwabara, Y. *Nippon Shokuhin Kogyo Gakkaishi* **1982**, *29*, 111-116.
3. Katayama, T.; Nagai, I. *Bulletin of the Japanese Society of Scientific Fisheries* **1960**, *26*, 29-32.
4. Huhtanen, C. N. *Journal of Food Protection* **1980**, *43*, 195-196.
5. Sekiyama, Y.; Mizukami, Y.; Takada, A.; Oosono, M.; Nishimura, T. *J. Antibact. Antifung. Agents* **1996**, *24*, 171-178.
6. Inoue, S.; Goi, H.; Miyauchi, K.; Muraki, S.; Ogihara, M.; Iwanami, Y. *J. Antibact. Antifung. Agents* **1983**, *11*, 609-615.
7. Tokuoka, K.; Isshiki, K. *Nippon Shokuhin Kogyo Gakkaishi* **1994**, *41*, 595-599.
8. Goi, H.; Inoue, S.; Iwanami, Y. *J. Antibact. Antifung. Agents* **1985**, *13*, 199-204.
9. Issiki, K.; Tokuoka, K.; Mori, R.; Chiba, S. *Biosci. Biotech. Biochem.* **1992**, *56*, 1476-1477.
10. Kanemaru, K.; Miyamoto, T. *Nippon Shokuhin Kogyo Gakkaishi* **1990**, *37*, 823-829.
11. Shofran, B. G.; Purrington, S. T.; Breidt, F.; Fleming, H. P. *Journal of Food Science* **1998**, *63*, 621-624.
12. Kojima, M.; Ogawa, K. *J. Ferment. Technol.* **1971**, *49*, 740-746.
13. Kleese, P.; Lukoschek, P.; *Arzneimittel-Forsch.* **1955**, *5*, 505-507.
14. Drobnica, L.; Zemanova, M.; Nemec, P.; Antos, K.; Kristian, P.; Stullerova, A.; Knoppova, V.: Nemec, P. Jr. *Applied Microbiology* **1967**, *15*, 701-709.
15. Drobnica, L.; Zemanova, M.; Nemec, P.; Kristian, P.; Antos, K.; Hulka, A. *Applied Microbiology* **1967**, *15*, 710-717.
16. Dornberger, K.; Boeckel, V.; Heyer, J.; Schoenfeld, CH.; Tonew, M.; Tonew, E. *Pharmazie* **1975**, *30*, 792-796.
17. Masuda, H.; Tsuda, T.; Tateba, H.; Mihara, S., Japan Patent 90,221,255, 1990.
18. Harada, Y.; Masuda, H.; Kameda, W., Japan Patent 95,215,931, 1995.
19. Masuda, H.; Harada, Y.; Tanaka, K.; Nakajima, M.; Tateba, H., In *Biotechnologyy for Improved Foods and Flavors*; Takeoka, G. R.; Teranishi, R.; Williams, P.; Kobayashi, A., Eds.; ACS Symposium Series 637; American Chemical Society: Washington, DC, 1996; pp 67-78.
20. Delaquis, P. J.; Mazza, G. *Food Technology* **1995**, *11*, 73-84.
21. Zsolnai, T. *Arnzeim. Forschung.* **1966**, *16*, 870-876.

22. Kawakishi, S.; Kaneko, T. *J. Agric. Food Chem.* **1987**, *35*, 85-88.
23. Banks, J. G.; Board, R. G.; Sparks, N. H. C. *Biotechnol. Appl. Biochem.* **1986**, *35*, 103-107.
24. Masuda, H.; Harada, Y.; Inoue, T.; Kishimito, N.; Tano, T., In *Flavor Chemistry of Ethnic Foods*; Shahidi, F.; Chi-Tang Ho, Eds.; Kluwer Academic/Plenum Publishers, New York, 1999, pp 85-96.
25. *Biopreservation*; Morichi, T.; Matsuda, T., Eds.; Saiwaishobo: Tokyo, Japan, 1999; pp 42-64.

Chapter 20

Processing Modulation of Soymilk Flavor Chemistry

Yu-Wen Feng, Terry E. Acree, and Edward H. Lavin

Department of Food Science and Technology, Cornell University, New York State Agricultural Experiment Station, Geneva, NY 14456

Undesirable beany odors in soymilk are mainly caused by enzymatic polyunsaturated lipid oxidation. Over the last 40 years many different chemical or physical processes were designed to eliminate the beany odor. This paper describes the use of Gas Chromatography-Olfactometry (GC-O) and Gas Chromatography Mass Spectrometry (GC-MS) to profile the aroma of soymilks made by aerobic and anaerobic processes. Aroma compounds like hexanal (grassy), 1-octen-3-one (mushroom), (E,Z)-2,6-nonadienal (cucumber), 2-nonenal (chalky), 2,4-decadienal (oily), methional (cooked potato), and beta-damascenone (floral) were measured. The processing effects will be discussed in terms of their impact on flavor.

Introduction

Soymilk is an aqueous extract of soybeans. This extract is called "milk" because the emulsion of protein and fat appear similar to cow's milk but the flavor is very different (*1,2*). The typical flavors of soymilk prepared by traditional East Asia methods are appreciated differently by people from Asian culture and patterns, but are described in pejorative terms, “beany,” “painty,” “rancid” and “bitter” by westerners. The positive descriptions of soymilk by westerners, “nutty,” “plain,” or “mild” are often not enough to overcome the "beany" flavor. Expanding the use of soymilk in the west is limited by the “beany” odor that usually dominates soymilk products that are not masked with flavor, like chocolate (*3*).

Research indicates that the odorant volatiles are formed enzymatically from mediated lipid oxidation through the action of lipoxygenase (*4*). Soybeans are approximately 20% lipid, 60% of which are polyunsaturated fatty acids including linoleic and linolenic acids with the (Z,Z)-1,4-pentadiene moiety. Off-flavors are not present in whole dry soybean, they formed as soon as the beans are soaked or ground with the smallest amount of water below 80°C. The lipoxygenase present in whole soybean acts upon those fatty acids containing a (Z,Z)-1,4-pentadiene moiety (*5*), and converts them to the corresponding acid with a 1-hydroperoxy-(E,Z)-2,4-pentadiene structure (*6, 7*). Hydroperoxide lyase catalyzes the development of volatiles with low molecular weights from the degradation of hydroperoxides (*8, 9*). The formation of flavor volatiles is also dependent on enal isomerases and alcohol dehydrogenase. Lipoxygenases can be inactivated with heat and acid (*10*).

In order to increase the acceptance of soymilk by westerners many methods have been developed to improve soymilk flavor. Because enzymatic polyunsaturated fatty acid oxidation is the major source of beany odor, we focused our efforts on controlling oxidation by limiting the concentration of oxygen. Native lipoxygenase enzyme is activated when a ferrous ion (+2) at the active site is oxidized to ferric ion (+3) by molecular oxygen or hydroperoxide. The active enzyme is able to react with a fatty acid to abstract hydrogen. When oxygen is sufficient, the reaction will follow this aerobic pathway to produce hydroperoxides from fatty acids. However, when the oxygen level is low, i.e. anaerobic, the radical enzyme-substrate complex will dissociate into a radical and the enzyme in the ferrous state. This catalyzes other hydroperoxides to decompose instead of producing more hydroperoxide from fatty acid. The decomposition of the hydroperoxides is the rate-limiting step of the anaerobic lipoxygenase reaction and the major products are n-pentane, oxo-dienoic acids, dimers, and water rather than the hydroperoxides produced during the aerobic reaction. These two pathways form different volatiles resulting in different aroma characters. Limiting the oxygen exposure will inhibit aerobic oxidation but not the anaerobic reaction. However, the anaerobic reaction needs some oxygen or hydroperoxide to activate the native enzyme. A small amount of oxygen will favor the anaerobic reaction; larger amounts favor the aerobic pathway (*11, 12*), Fig. 1. It would be possible to produce soymilk without the beany odor if we could eliminate oxygen exposure during the entire process.

Gas chromatographic studies of soymilk volatiles conducted in the 1960s and 1970s (*14, 15, 16*) revealed most of the odor active volatiles. The use of gas chromatography-mass spectrometry (GC-MS) has revealed many more volatile products of lipid oxidation but which ones contribute to soymilk flavor is still not clear. This study used gas chromatography – olfactometry (GC-O) to investigate which compounds are the most potent in soymilks made by aerobic and anaerobic processes.

Experimental

Materials & Equipment

Canadian Grade No. 1 soybeans were stored at 0° C until needed. A stainless steel Waring Blender (1 L) was used for grinding the soybeans. A water bath (Precision Scientific Co., Chicago, IL) was used for the heat treatment to make traditional soymilk and pasteurized soymilk. Coarsely woven cheesecloth was used for the filtration of the soymilk slurry.

A glove box (Allied Engineering & Production Crop, Alameda, CA), was used for processing. An oxygen monitor GC-501, range: 0-25 % (G.C. Industries, Inc, Fremont, CA) was used to measure the oxygen concentration in the atmosphere of the glove box. YSI model 50B dissolved oxygen meter was used for measuring the dissolved oxygen residual in the water flushed and deoxygenated with high-purity nitrogen.

Gas Chromatography Olfactometry (GC-O)

CharmAnalysis™ was accomplished by a Hewlett Packard 5890 gas chromatograph equipped with a 12m x 0.32 mm cross-linked methyl silicone fused silica capillary column (film thickness= 0.33 µm). The effluent consisting of helium (2 ml /min) as the carrier gas and nitrogen as the make up gas (ca. 30 ml /min), was mixed with the sniffing air (20 L/min) which was 99% laboratory air humidified to between 50% and 75% humidity, passed to the sniffer through a 10mm diameter silylated pyrex tube.

Gas Chromatography-Mass Spectrometry(GC-MS)

Mass spectrometric identifications were performed using a Hewlett Packard 5890 GC/MS with an ion source by electron impact (EI) data system equipped with a 25m x 0.20mm cross-linked methyl silicone fused silica capillary column (film thickness=0.33µm).

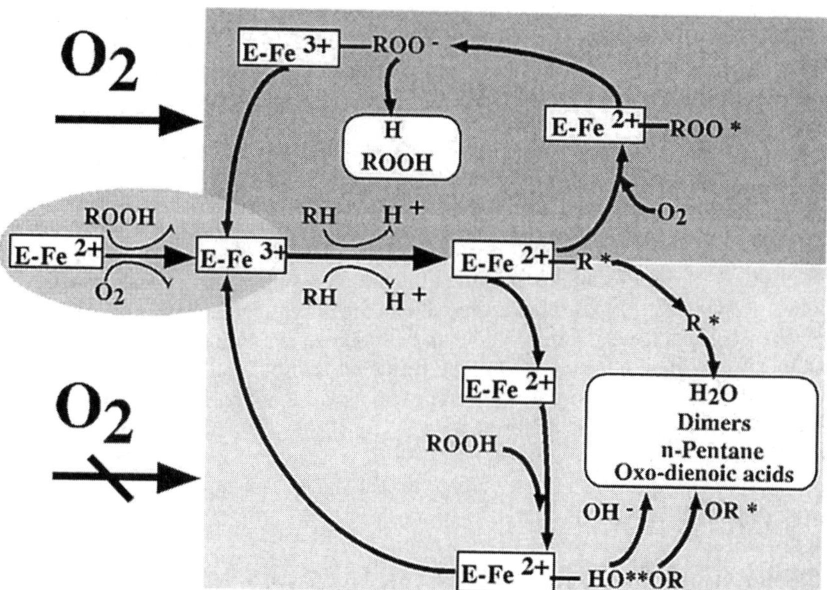

Figure 1. The proposed mechanism of aerobic (top scheme), and anaerobic (bottom scheme) lipid oxidation by lipoxygenase. (13)

Soymilk preparation

Aerobic Soymilk Process

Soybeans were cleaned and soaked in five times their weight of distilled water at 25°C for about eight hours. After discarding the soaking water, soybeans weighing twice their original weight were drained and blended with two times their dry weight of distilled water for about five minutes and then blended with an additional seven times their dry weight of distilled water for about two minutes, at 20°C. The slurry (water added: dry weight of soybeans was 9:1) was simmered at 95 to 98°C for 20 minutes in a water bath, and then filtered through a coarsely woven filtercloth and squeezed by hand to express as much soymilk as possible. The soymilk was pasteurized by simmering in a water bath at 85 to 90°C for 10 minutes to reduce microbial contamination and stored at 4° C overnight before extraction the next day.

Anaerobic Soymilk Process

Distilled water was flushed with nitrogen until the dissolved oxygen was below 5% of the original concentration. Soybeans were placed into a vacuum foil. The blender, a stir bar, a heating flask with a reflux condenser and a T-shaped stopper were placed in the glove box. The oxygen concentration in the glove box was reduced to 0.2-2% by purging with nitrogen. The same procedure as previously described for the production of soymilk was carried out inside the glove box: soybeans were cleaned, soaked drained and blended to a slurry. The slurry (water added: dry weight of soybeans was 9:1) was transferred to a flask provided with a head-space of nitrogen. The following procedures were done outside the glove box: the slurry was heated and stirred at 95 to 98°C for 20 minutes while the headspace of the slurry was purged with nitrogen (Fig 2). After the heat treatment, the slurry was filtered and pasteurized in as previously described for the aerobic soymilk process (*1*).

Extraction and Concentration

Soymilk was extracted with a 0.67 portion of Freon™ 113 for at least 30 minutes. After removal of the Freon™ 113 extract the aqueous phase was extracted with a 0.67 portion of ethyl acetate. After the collection of the ethyl acetate extract the aqueous phase was discarded. Every Freon™ 113 and ethyl acetate extract was filtered through magnesium sulfate to remove as much water as possible and concentrated to 1ml using a Buchi 0.1 rotary evaporator. Freon extracts were evaporated under 48 kilopascal (kPa.) and ethyl acetate under 86 kPa.

GC-O and GC-MS Analysis

GC-O Dilution Analysis

Extracts of samples were concentrated to 1 ml and a series of 1:3 dilutions were prepared for GC-O analysis. Samples at the highest concentration were used for GC-MS analysis. One µl of each concentration was injected into a modified Hewlett Packard (Palo Alto, CA) 5890 gas chromatograph equipped with a 12 m

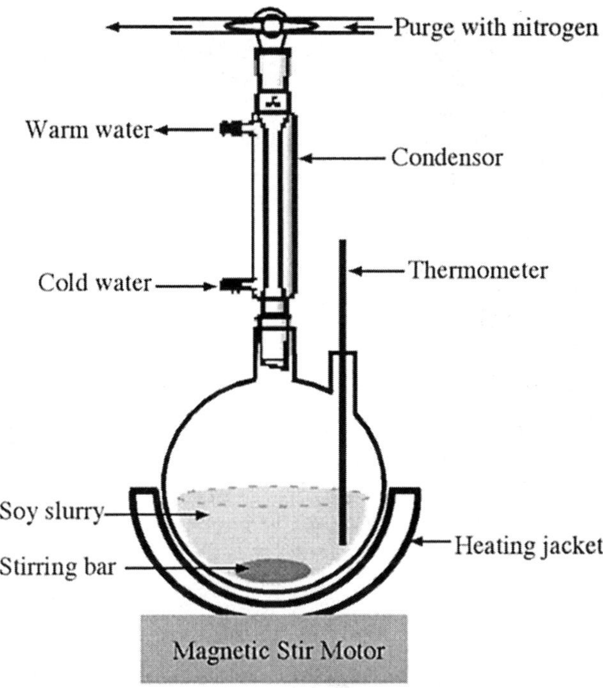

Figure 2. Heating apparatus for soy slurry under anaerobic conditions.

x 0.32 mm cross-linked OV101 fused silica column. The oven temperature was held at 35° C for three minutes and then programmed to 225° C at 6° C/min. The injector temperature was 200° C and the detector was held at 225° C. Retention times of all odor active compounds were recorded on a Macintosh™ computer and converted to retention indices by linear interpolation of retention times of a series of 7-18 carbon alkane standards run under identical conditions and detected with a flame ionization detector (FID). The GC-O chromatograms were constructed using Charmware software (*17*). All GC-O runs and all samples were replicated. A sniffer without specific anosmias screened by a standard solution should be able to theoretically detect those compounds by GC-O (*18,19*). A standard solution consisting of eight compounds: ethyl butyrate 600 ng/µl, ethyl hexanoate 720 ng/µl, 1,8 cineole 600 ng/µl, carvone 300 ng/µl, ethyl 2-methyl butyrate 100 ng/µl, methyl anthranilate 300 ng/µl, ß-damascenone 2.5 ng/µl, and o-aminoacetophenone 100 ng/µl was used to test the sniffer's acuity. In this study a single sniffer was used. All of the GC/O analyses were done in a room free of competing odorants. The room was an enclosed facility with filtered air entering through the top of the east wall and exiting through the bottom of the south wall. Composite charm chromatograms of both samples and the peaks representing the odor active compounds were resolved and integrated to give charm values using software sold by DATU Inc, Geneva, NY.

Identification of Odor Active Compounds

Freon™113 and ethyl acetate extracts from one replicate were analyzed by GC-MS. 1μl was injected into the GC-MS which had a starting temperature of 35°C increasing after three minutes at 4°C/min to 225°C. The injector temperature was 200°C, the detector was held at 225° C. Compounds analyzed in the extracts were those within five RI of an odor detected during GC-O. Identification of all compounds was confirmed by a retention index and spectrum match with that of an authentic standard run under identical conditions. Odor description and retention index were used as final identification.

Data presentation

The data in this study will be presented by GC-FID chromatograms, Charm chromatograms and odor spectra. In a Charm chromatogram a peak is constructed from the on-off responses recorded in the computer while the sniffer is sniffing at the GC-O by using CharmAnalysis. Charm values are the peak areas generated by combining the on-off data as a function of retention index. The height of these peaks is the FD value for the compound and their area is their Charm values (20).

An odor spectrum is based on the idea of Steven's Law, $\Psi = k\Phi^n$, where Ψ is the perceived intensity of a stimulant, k is a constant, Φ is the stimulus level and n is an exponent between 0.3 to 0.8 for odor. Using $n = 0.5$, the median value for olfaction, to calculate the odor spectrum value (OSV), Ψ, and a measure of odor potency as Φ an odor spectrum can be generated. An odor spectrum chromatogram is a plot of odor spectrum values against retention indices, and shows the pattern of relative potency independent of concentration (*21, 22*). The reason to use OSV is that GCO data expressed as FD values or charm values have no absolute meaning. They are relative measures of odor potency as the odorants elute from a GC. Furthermore, the relative concentrations of the volatiles in the samples injected in the GCO are biased by the method used to prepare the sample. However, when someone sniffs a mixture of volatiles in the headspace above a food (orthonasal smell) they are experiencing the volatiles at different relative potencies than when they sniffed extracts eluting from a GCO (*21*). This difference between the GCO data from extracts and the composition of orthonasal headspace is even greater with foods containing fats such as soymilk (*20*). Therefore, using OSV to express the sniffing results might give a better understanding of the aroma perception of soymilk.

Results

Aroma from aerobically processed soymilk

The 15 most odor potent compounds found in soymilk made by a typical aerobic process are presented in Figure 3. The GC-FID chromatogram, the Charm Chromatogram made by GC-O dilution analysis and the odor spectrum of soymilk are also shown. The most odor potent compounds as shown in Table I were lipid oxidation

Glycine max-Aerobic process

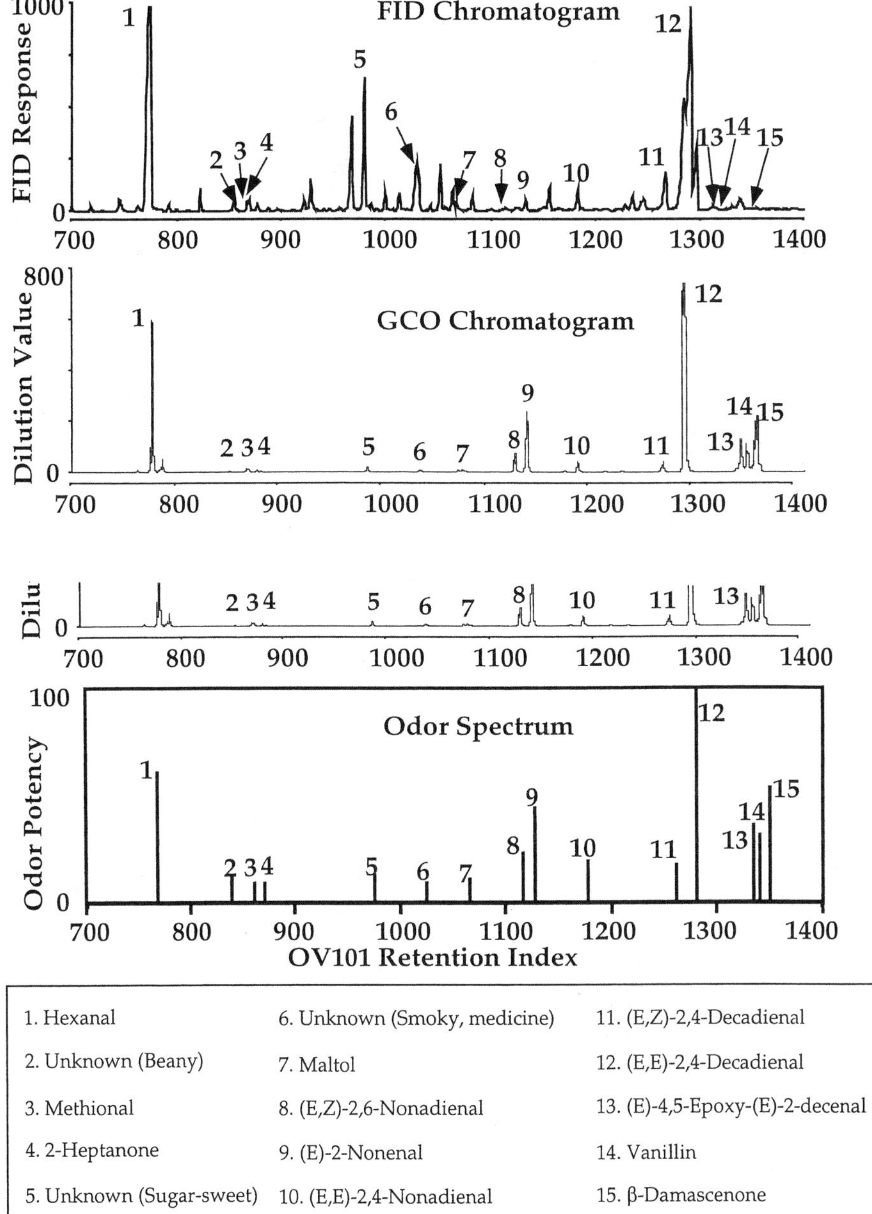

Figure 3. GC-FID chromatogram, charm chromatogram, and odor spectrum of soymilk made by aerobic process.

Table I. Odor spectrum value of soymilk made by aerobic process

Odor spectra	*Compound*	*Retention Index (OV101)*	*Odor Spectrum Value (OSV)*
Grassy	Hexanal	768	61
Wheaty	Unknown	840	13
Rancid	Methional	862	9
Beany	2-Heptanone	871	9
Sugar -sweet	Unknown	976	13
Smoky	Unknown	1026	9
Syrup-sweet	Maltol	1067	11
Cucumber	(E,Z)-2,6-Nonadienal	1117	23
Chalky	(E)-2-Nonenal	1129	45
Off-oil	(E,E)-2,4-Nonadienal	1179	20
Off-oil	(E,Z)-2,4-Decadienal	1263	18
Off-oil	(E,E)-2,4-Decadienal	1283	100
Metallic	(E)-4,5-Epoxy-(E)-2-decenal	1337	36
Cream	Vanillin	1342	32
Flower-sweet	β-damascenone	1352	54

products: (E,E)-2,4 decadienal and n-hexanal. Methional, maltol, (E,Z)-2,6-nonadienal, (E)-4,5-epoxy-(E)-2-decenal, vanillin and β-damascenone, detected in this study, have not been previously reported in soymilk (*15, 23*).

In this study, GC/O analysis, focusing on the odor active volatiles, detected compounds with very low odor thresholds and low amounts in soymilk, such as β-damascenone (odor threshold 2-20pg/g in water) (*23*), (E)-4,5-epoxy-(E)-2-decenal (odor threshold 5ppb in water) (*22*), (E,Z)-2,6-nonadienal that were not detected in previous GC/MS studies. Research that did not follow the heating treatment of the traditional Asian process, similar to the aerobic process mentioned in this study, involving soaking, blending, and cooking, to inactivate the trypsin inhibitor, were not able to detect methional, β-damascenone, and maltol. The use of solvent extraction in this study prevented volatiles with low molecular weight (with Retention Index<700 on OV101) from detection, therefore, over emphasizing the odor potency of some higher molecular weight odorants (*21*).

Aroma from Anaerobically Processed Soymilk

Figure 4 shows the GC-FID chromatogram, the Charm chromatogram made by GC-O dilution analysis sniffing and the odor spectrum of anaerobically produced soymilk. The odor spectrum values in Table II show that no n-hexanal was detected by GC-O sniffing but β-damascenone, known to be increased by thermal degradation of its precursors, was the most odor potent compound. 4-Vinyl-2-methoxyphenol, a thermal degradation product of phenolic compounds, and 1-octen-3-one, a compound with a mushroom note were among the most odor potent compounds. 1-Octen-3-ol, a major peak in the GC-FID chromatogram, however, had very low potency in the GC-O study. There were still some lipid oxidation products found in anaerobically produced soymilk because of lipid autoxidation occurred during the post harvest handling of soybeans. Hydroperoxides produced during postharvest initiated some lipid oxidation reaction. The "anaerobic condition" in this study was not absolutely anaerobic. However, the disappearance of n-hexanal from the GC-O sniffing data showed that the aerobic oxidation was under control.

Comparison of aerobic and anaerobic processes

Figure 5 shows the comparison of the two treatments. The GC-FID chromatograms show that the volatile content was decreased significantly within the sample processed under anaerobic conditions. The Charm chromatograms showed the odor intensity of soymilk made anaerobically was weaker than that made in the regular atmosphere. In the odor spectra the normalized data present the similarity or the difference of the odor patterns of these two soymilks.

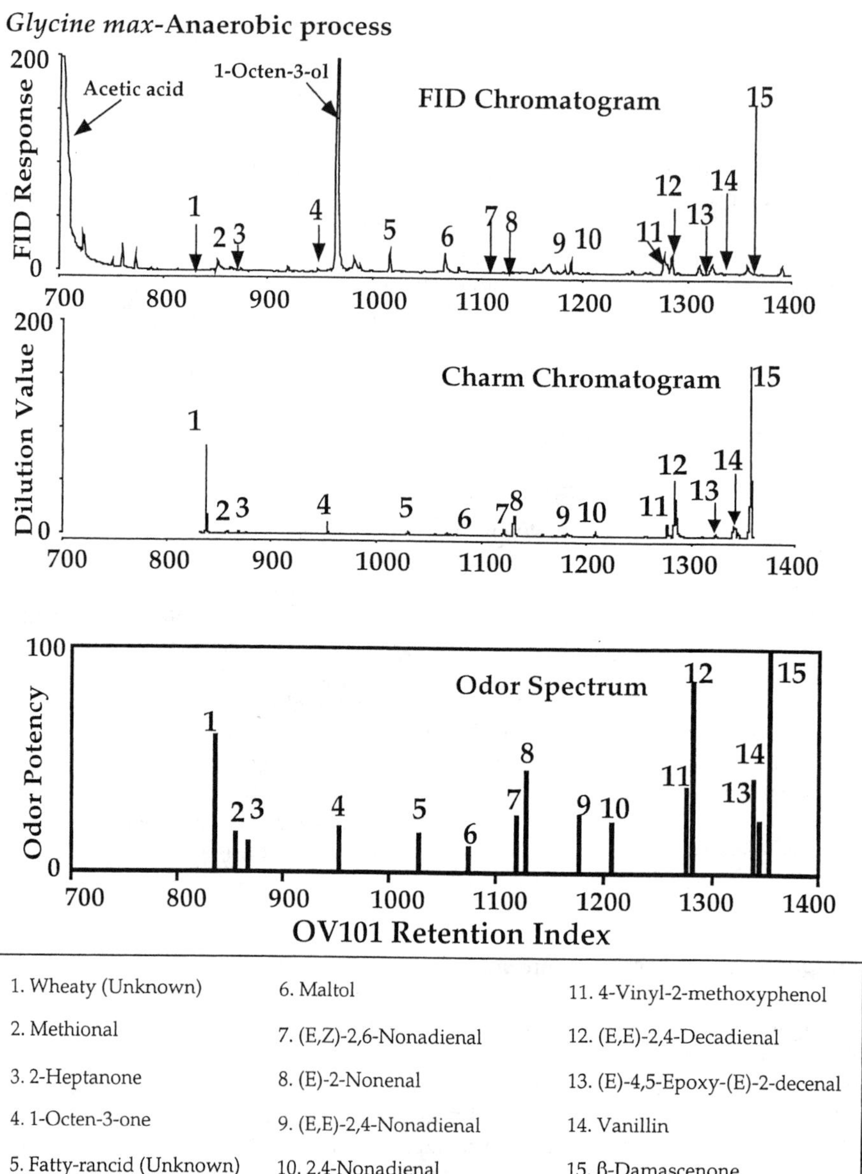

Figure 4. GC-FID chromatogram, charm chromatogram, and odor spectrum of soymilk made by anaerobic process.

Table II. Odor spectrum value of soymilk made by anaerobic process

Odor spectra	*Compound*	*Retention Index (OV101)*	*Odor Spectrum Value (OSV)*
Wheaty	Unknown	840	61
Rancid	Methional	862	17
Beany	2-Heptanone	871	14
Mushroom	1-Octen-3-one	953	21
Fatty-Rancid	Unknown	1030	18
Syrup-sweet	Maltol	1067	12
Cucumber	(E,Z)-2,6-Nonadienal	1117	26
Chalky	(E)-2-Nonenal	1129	46
Off-oil	(E,E)-2,4-Nonadienal	1179	26
Off-oil	2,4-Nonadienal	1207	23
Band-aid	4-Vinyl-2-methoxyphenol	1276	39
Off-oil	(E,E)-2,4-Decadienal	1283	86
Metallic	(E)-4,5-Epoxy-(E)-2-decenal	1337	43
Cream	Vanillin	1342	24
Flower-sweet	β-Damascenone	1352	100

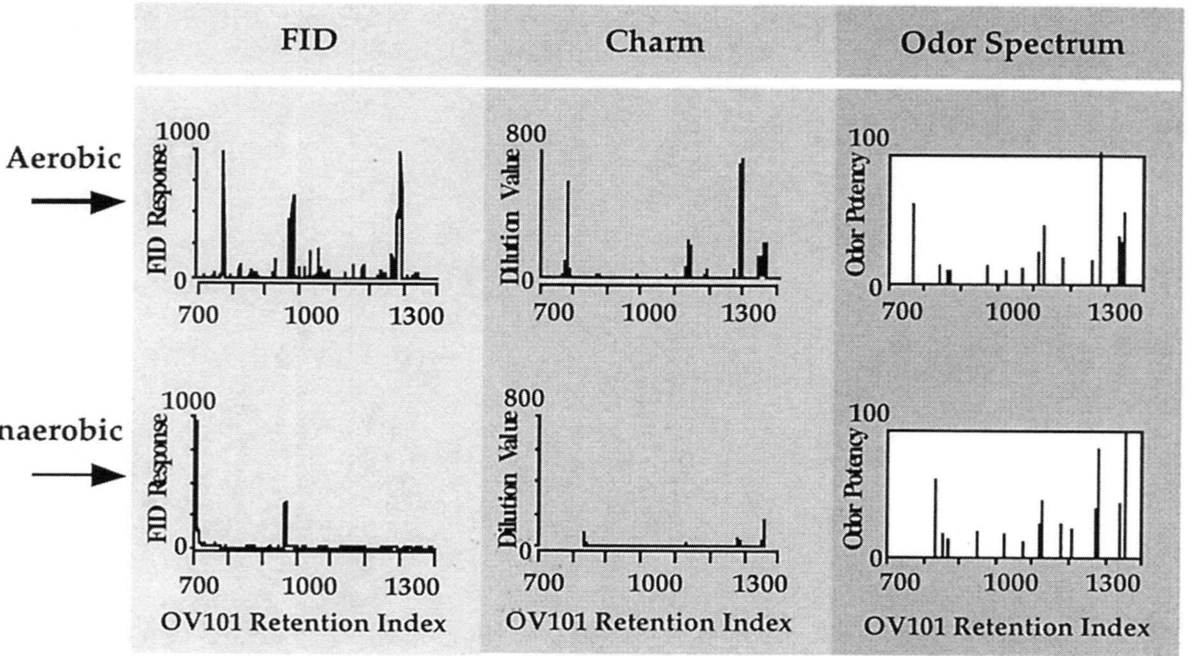

Figure 5. The comparison of aerobic (above) and anaerobic (below) processing by GC-FID chromatograms (left), charm chromatogram (middle), and odor spectra (right).

Discussion

Polyunsaturated lipid oxidation and thermal reactions play major roles in the formation of odorant volatiles in soymilk. In this study controlling the atmosphere oxygen provided a possibility to control the formation of the beany odor. However, in order to inactivate enzymes and some anti-nutrition factors, a certain time at the high temperature (around the boiling point) is required (*24, 25*). Therefore, in addition to controlling enzymatic lipid oxidation during soymilk manufacture the impact of heating on the aroma of soymilk in terms of both aerobic and anaerobic processing need more investigation. Controlling lipid autoxidation during postharvest handling of soybeans may enhance the aroma anaerobically and aerobically.

Literature Cited

1. Lo, W. Y., M.S. thesis, Cornell University, Ithaca, New York, U.S.A. 1967.
2. Smith, A. K.; Beckel, A. C. *Chem. Eng. News ACS.* **1946,** 24, 54-56.
3. Shurtleff, W.; Aoyagi, A. *Tofu and Soymilk Production.* 1st ed. 1979, New-Age Food Study Center: Lafayette, NY, 1979; pp 197-198.
4. Wilkens, W. F.; Mattick, L. R.; Hand, D. B. *Food Tech.* **1967,** 21, 86.
5. Gardner, H. W. In *Flavor Chemistry of Fats and Oils;* Min, D. B. and Smouse, T. H. Ed.; American Oil Chemists' Society: U.S.A., 1985; pp 447-504.
6. Goodwin, T. W.; Mercer, E. I. *Introduction to Plant Biochemistry*; Pergamon Press Ltd.: New York, 1983.
7. Schewe, T.; Rapoport, S. M.; Kuhn, H. *Advanced Enzymology.* **1986,** 58, 191-273.
8. Zimmermann, D. C. *Biochem. Biophys. Res. Commun.* **1966,** 23, 398.
9. Gini, B.; Koch, R. B. *J. Food. Sci.* **1961,** 26, 359.
10. Farkas, D. F.; Goldblith, S. *J. Food. Sci.* **1961,** 27, 262.
11. Chism, G. W. In *Flavor Chemistry of Fat and Oils;* Min, D. B. and Smouse, T. H. Ed.; American Oil Chemists' Society: U.S.A., 1985; pp 175-187.
12. Hsieh, R. J. In *Lipid in Food Flavors;* Ho, C. T.; Hartman. T. G., Ed. American Chemistry Society: Washington, D.C., 1994, pp 30-48.
13. Mattick, L. R.; Hand, D. B. *Agric. Food. Chem.* **1969,** 17**,** 15.
14. de Groot, I. M. C.; Veldink, G. A.; Vliegenthart, J. F. G.; Boldingh, J. ; Wever, R.; Van Gelder, B. F. *Biochim. Biophys. Acta.* **1975,** 377, 71.
15. Wilkens, W. F.; Lin, F. M. *J. Agr ic. Food. Chem.*, **1970,** 18(3), 337.
16. Lao, T. B. Ph.D. thesis, University of Illinois, Illinois, 1971.
17. Acree, T. E.; Bernard, J.; Cunningham, D. G. *Food Chemistry.* **1984,** 14, 273-286.
18. Amoore, J. E., *Chemical Senses and Flavor*, **1979,** 4(2), 153-161.
19. Marin, A. B.; Acree, T. E.; Barnard, J. *Chemical Senses.* **1988,** 13(3), 435-444.

20. Feng, Y.-W.; Acree, T. E. *Foods Food Ingredients J. Jpn.* **1999,** 179, 57-66.
21. Roberts, D. D. Ph.D. thesis, Cornell University, Ithaca, New York, 1996.
22. Ong, P. K. C.; Acree, T. E.; Lavin, E. H. *J. Agric. Food. Chem.* **1998,** 46, 611-615.
23. Kobayashi, A.; Tsuda, Y.; Hirata, N.; Kubota, K.; Kitamura, K. *J. Agric. Food Chem.* **1995,** 43, 2449-2452.
24. Ohloff, G. Progress Chem. Org. Natural Products. **1978,** 35, 431-527.
25. Sangle, M.; Devi, R.; Pawar, V. D.; Arya. *Journal of Dairying Foods & Home Sciences.* **1993,** 12,113-119.
26. Friedman, M.; Brandon, D. L.; Bates, A. H.; Hymowitz, T. *J. Agric. Food Chem,* **1991,** 39(2), 327-335.

Indexes

Author Index

Subject Index

G

H

I

K

L

M

N

P

S

Y